I0074367

Combinatorics Problems and Solutions, Second Edition
by Stefan Hollos and J. Richard Hollos

Abrazol Publishing
an imprint of Exstrom Laboratories LLC
662 Nelson Park Drive, Longmont, CO 80503-7674 U.S.A.

Publisher's Cataloging in Publication Data
Hollos, Stefan
Combinatorics Problems and Solutions, Second Edition / by Stefan Hollos and J. Richard Hollos
p. cm.
ISBN: 978-1-887187-48-0
Library of Congress Control Number: 2024934303
1. Combinatorial analysis–Problems, exercises, etc. 2. Combinatorial enumeration problems
I. Title. II. Hollos, Stefan.
QA164.8 .H65567 2024
511.62 HOL

Cover image: Paul Cezanne - Still Life with Cherries and Peaches, 1885-1887, https://commons.wikimedia.org/wiki/
File:Paul_Cezanne_-_Still_Life_with_Cherries_and_Peaches,_1885-1887.jpg

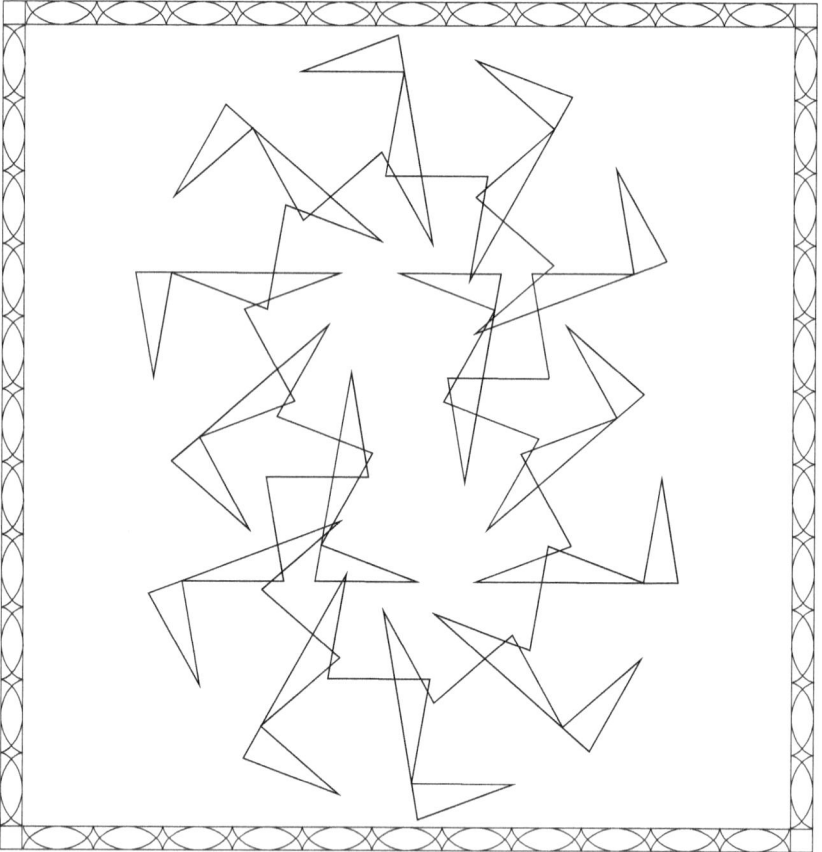

*Study and in general the pursuit of truth and beauty is
a sphere of activity in which we are permitted to
remain children all our lives.*
Albert Einstein

Contents

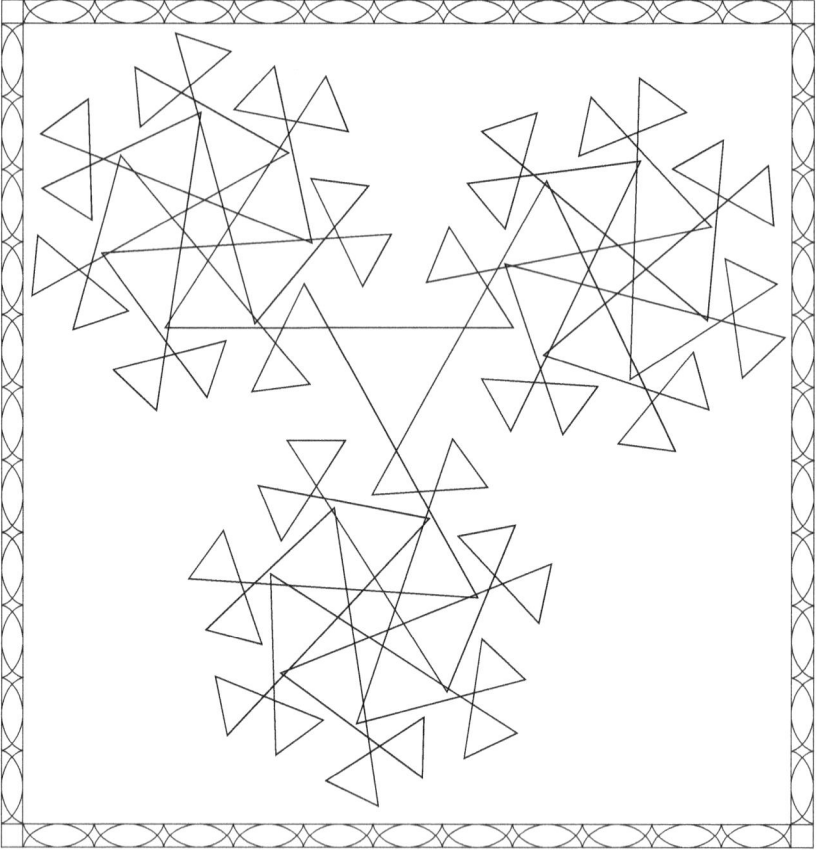

It is not the possession of truth, but the success which attends the seeking after it, that enriches the seeker and brings happiness to him.

Max Planck

In this new edition we have expanded the introductory section by more than twice the original size, and the number of problems has grown by over 30%. There are new sections on the pigeon hole principle and integer partitions with accompanying problems. Many of the new problems are application oriented. There are also new combinatorial geometry problems.

Someone with no prior exposure to combinatorics will find enough introductory material to quickly get a grasp of what combinatorics is all about and acquire the confidence to start tackling problems.

In the first edition we had 2 separate sets of problems. In this new edition, we have combined the two sets, and removed some of the easier problems. The last 60 problems at the end are all new, with a few added also in between.

Combinatorics has so many interesting applications that we could not possibly cover them all in this book. One area of great interest to us is the application of combinatorics to physics. When we find interesting new combinatorics problems we will post them on our blog at:

https://exstrom.com/blog/abrazolica/

A book like this can never be truly finished. The subject is too broad and deep. Our plan is to provide collections of new problems on the book's website at:

https://www.abrazol.com/books/combinatorics1e2/

We hope you enjoy the intellectual adventure you are about to embark on by reading this book.

We can be reached by email at:

stefan[at]exstrom DOT com
richard[at]exstrom DOT com

You can sign up for our newsletter at:

https://www.abrazol.com

Stefan Hollos and J. Richard Hollos
Exstrom Laboratories LLC
Longmont, Colorado, U.S.A.
March 2024

Why would anyone want to solve combinatorics problems? The best reason has to be because it's simply fun. It's also a great way to sharpen your problem solving skills and give your brain a good workout. An hour spent solving combinatorics problems is better than an hour spent playing chess since it's not only fun but it gives you skills you can use in mathematics, computer science, physics, biology, etc.

Let's define what we mean by combinatorics. Much of combinatorics is based on the idea of counting. This may involve simply counting the number of elements in a set. It sounds simple but it's not a matter of just looking and counting. The set is usually only defined as elements that meet some condition or have some property. You could, in principle, construct the set from the definition and then count the number of elements it contains. When the number of elements runs into the thousands, millions, billions, or becomes infinite then things become quite tedious. So to solve combinatorics problems you often need some insight, ingenuity, and the ability to turn the problem into a different form that is more easily solved. What could be more fun than that?

This is not meant to be a textbook on combinatorics but there is enough introductory material so that even

3

someone with little or no prior exposure to the subject can get something out of it. Some familiarity with the concept of sets, subsets, factorials, and basic algebra is all that is required. We start with some definitions in the introduction along with a guide to solving counting problems. The guide is a list of some of the most common counting problems. There is an equation for each problem and a set of equivalent descriptions of the problem. This is followed by short sections that explain some of the most basic principles in combinatorics. They were chosen because they are used in the problems but they are by no means exhaustive.

After the introduction comes the problems and exercises. The problems are generally easier and shorter than the exercises and increase in complexity as you go. Some of the exercises can be quite involved and may take some time to fully work out. A computer algebra system capable of dealing with very large numbers may be helpful for some of the problems and exercises (Emacs calc also works). Each problem and exercise is fully worked out in detail. Most of the problems and exercises are modernized versions of the problems and exercises found in the book: *Choice and Chance* by W. A. Whitworth (see Further Reading at the end of the book). Happy problem solving.

We can be reached by email at:
stefan[at]exstrom DOT com
richard[at]exstrom DOT com

Stefan Hollos and J. Richard Hollos

Exstrom.com

QuantWolf.com

Exstrom Laboratories LLC

Longmont, Colorado, U.S.A.

January 2013

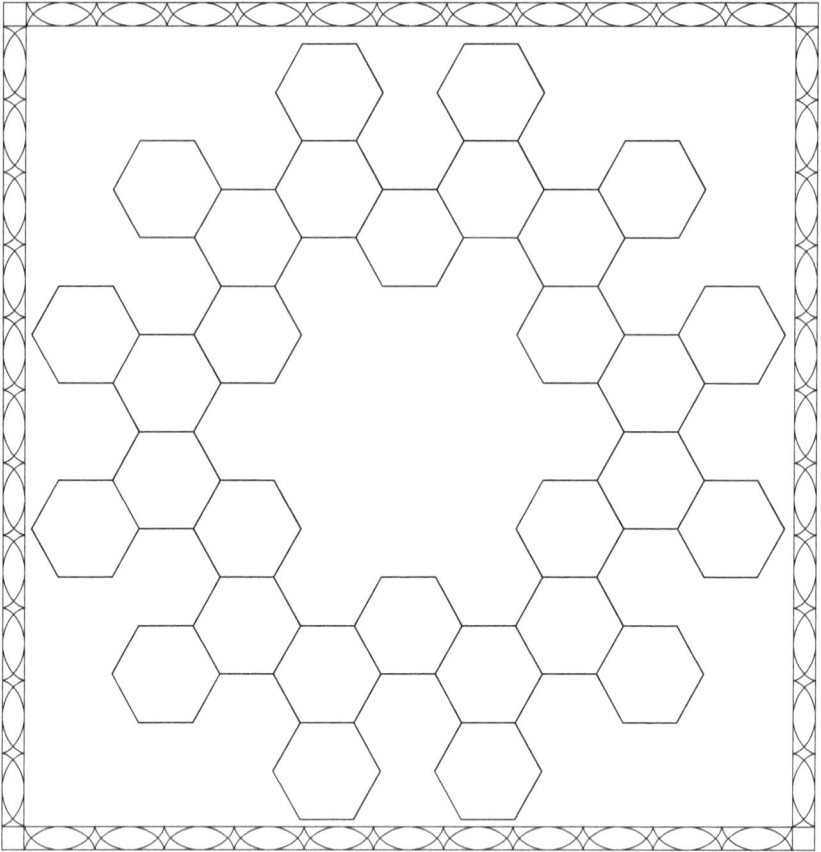

Mathematics is one of the essential emanations of the human spirit, a thing to be valued in and for itself, like art or poetry.

Oswald Veblen

The following is a short and simple introduction to some of the most elementary concepts in enumerative combinatorics. It assumes almost no mathematical background beyond basic algebra. In addition to being an introduction to combinatorics, it provides good background material for solving problems in discrete probability.

Definitions

We start with a short list of definitions so that you know what we're talking about. You probably know this stuff already but it's a good idea to skim over it to avoid any possible confusion and to make sure we're both talking the same language.

- A set is a collection of objects called elements or members. All the elements of a set are unique. The elements of a set are not ordered. The set of the first four letters of the English alphabet can be written as $\{a, b, c, d\}$, or $\{d, c, b, a\}$, or any other order. The number of elements in a set is called its size or cardinality.

- The set B is a subset of the set A if every ele-

7

ment of B is also an element of A. The empty set (a set with no elements) is a subset of every set. A subset can be constructed as an unordered sampling of elements from a set.

- The set $[n] = \{1, 2, 3, \ldots, n\}$ is a subset of the set of integers.

- A multiset is a collection of objects where each object may occur one or more times. $\{a, a, a, b, c, c, d\}$ is a multiset with 3 copies of element a, 1 copy of b, 2 copies of c, and 1 of d. Like a set, a multiset is not ordered. A multiset can be constructed as an unordered sampling with replacement from a set.

- A list is an ordered collection of elements. The list $[a, b, c, d]$, differs from the list $[d, c, b, a]$. A list may also be called a permutation of the elements it contains. A list can be constructed as an ordered sampling of elements from a set.

- A circular list is an ordered collection of elements. The order must be unique with respect to a circular shift of the elements. The lists $[a, b, c]$, $[b, c, a]$, and $[c, a, b]$ are all the same circular list. A circular list may also be called a circular permutation of the elements it contains.

- An alphabet is a set of symbols also called letters.

- A multi-alphabet is a multiset of symbols, i.e. it may contain more than one copy of any or all of the symbols.

- A word is a list of symbols usually drawn from a specified alphabet or multi-alphabet. The number of symbols in a word is called its length or size.

- A partition of the set A into k parts is a set of k subsets of A with each element of A appearing in exactly one of the subsets and none of the subsets equal to the empty set.

- A composition of the integer n into k parts is a sum of k integers that equals n. The integers must be positive and may not equal zero. The order of the integers matters so that $2 + 3$ and $3 + 2$ are two different compositions of 5 into 2 parts.

- A weak composition of an integer is a composition where some of the integers may be zero.

- A partition of the integer n into k parts is a sum of k integers that equals n. The integers in the sum are called the parts of the partition. They must be positive and may not equal zero. The order of the integers does not matter so that $2+3$ and $3+2$ are the same partition of 5 into 2 parts.

- The partition number $p(n, k)$ is the number of ways to partition the integer n into k parts. For example $p(6, 3) = 3$ and the partitions are: $1 + 1 + 4$, $1 + 2 + 3$, $2 + 2 + 2$.

- The partition number $p(n)$ is the number of ways to partition the integer n into any number of parts.

$$p(n) = \sum_{k=1}^{n} p(n, k)$$

For example $p(4) = 5$ and the partitions are: $1 + 1 + 1 + 1$, $1 + 1 + 2$, $1 + 3$, $2 + 2$, 4.

- The Stirling number of the second kind $S(n, m)$ is the number of ways to partition a set of size n into m nonempty subsets. It is defined as follows:

$$S(n, m) = \frac{1}{m!} \sum_{k=0}^{m} (-1)^k \binom{m}{k} (m - k)^n$$

where $\sum_{k=0}^{m}$ means to sum what follows from $k = 0$ to m. The remaining notation in this expression is explained in the section Permutations and Combinations.

Set Enumeration

Much of combinatorics involves counting the number of elements in a set and the absolute simplest way to do that is by enumeration. This is the form of counting that every child first learns. You take the elements of the set one by one and assign integers to them starting with 1. The largest integer used is the number of elements in the set. This is called the size of the set or its cardinality. For example the set of odd decimal digits is $\{1, 3, 5, 7, 9\}$. The size of this set is 5.

Let's say you want to snack on some fruit. You look in the fridge and find the following set of fruits: $\{apple, orange, banana, mango\}$.
There are 4 fruits in this set so there are 4 ways you can have a snack. You probably get the idea behind simple enumeration.

If there were nothing more to combinatorics then it would be time to move on to something else. Unfortunately it's not always so simple. The elements of a set are often just defined in terms of some rule or some method of construction and there could be millions, billions or an infinite number of them. Take for example the set of all possible ways that 5 cards can be shuffled. There are 120 ways to shuffle them and you could in principle list all the ways and then count them but this is tedious to say the least. In general it's not practical to just list the elements of a set and then count them. It takes more ingenuity than that.

Sum Rule

The next step up in counting sophistication is called the sum rule. Here instead of just one set, we have two or more. In how many ways can an element be selected from one of the sets? For example if we have the set of 26 letters in the English language and the set of 10 decimal digits, how many ways can a letter or a digit be selected? There are 26 choices for the letter and another 10 choices for the digit so the total number of choices is $26 + 10 = 36$.

Going back to the snacks. Let's say that in addition to fruit you are considering a salty snack. In the cupboard you find the following set of snacks:
$\{chips, popcorn, pretzels\}$.
These are three more choices you have in addition to the four fruits so the total number of choices is $4 + 3 = 7$. The assumption here is that you will have a fruit or a salty snack but not both. We will get to the number of ways of having both in the next section.

To sum it up, the sum rule says that if you have two or more sets then the number of ways of selecting an element from one or the other of the sets is the sum of the sizes of the sets. For k sets with sizes n_1 through n_k the number of ways of selecting an element from any one of the sets is

$$N = n_1 + n_2 + \cdots + n_k \tag{1}$$

It seems like a very simple rule but there is one thing you have to be careful about when applying it. There must be no overlap between the sets. In other words the same element cannot appear in more than one set.

Another way of saying this is that the intersection of the sets must be the empty or null set. If there is a nonempty intersection then you have to take this into account. How to do that is covered in the section on the inclusion-exclusion principle below. There may however be situations where the same element appears in more than one set for a reason and each appearance has to be counted. It depends on the problem so one has to be careful.

Product Rule

Now suppose we have two or more sets and the question is how many ways can one element be selected from each of the sets. Let's return to the snacks. Now we're really hungry so we decide to have both a fruit and a salty snack. What are the possibilities? If we go for an apple then we can still pick any of the three salty snacks so there are three possibilities in this case. The same is true for the orange, banana, and mango. Each can be paired up with one of the three salty snacks. For each of the four fruits there are three possible salty snacks so the total number of ways of choosing a fruit and a salty snack is $4 * 3 = 12$. They are all listed below.

(apple, chips)	(apple, popcorn)
(apple, pretzels)	(orange, chips)
(orange, popcorn)	(orange, pretzels)
(banana, chips)	(banana, popcorn)
(banana, pretzels)	(mango, chips)
(mango, popcorn)	(mango, pretzels)

So for two sets you multiply their sizes to get the number of ways of making a selection of one element from each set. But what if we're thirsty and need to make a choice from a set of drinks {water, soda, juice, coffee, tea, beer}. This is a third set of six elements to choose from. For every one of the above combinations of snacks we can choose one of these six drinks, so the total number of possible snacks and drinks is $4 * 3 * 6 = 72$. That's quite a variety to choose from.

The general pattern should be clear. If you have k sets with sizes n_1 through n_k then the number of ways of selecting one element from each of the sets is:

$$N = n_1 * n_2 * \cdots * n_k \tag{2}$$

We will call this the product rule.

The sum and product rules are sometimes used together. Let's say you have three sets A, B, and C with sizes n_A, n_B, and n_C. How many ways can you select two elements from two different sets? The product rule says that the number of ways of selecting one element from A and one from B is $n_A * n_B$. This is one set of possibilities. You can also select one element from A and one from C in $n_A * n_C$ ways. This is another set of possibilities. Finally you can make one selection from B and one from C in $n_B * n_C$ ways. You have three sets of ways to make the selection and you have to pick from one of them so the sum rule says the total number of ways to make the selection is

$$N = n_A * n_B + n_A * n_C + n_B * n_C \qquad (3)$$

Permutations

Instead of selecting just one element from a set suppose we want to select k elements. How many ways can this be done? If the set has size n then there are n choices for the first element. The second choice is made from the remaining $n-1$ elements, so there are $n-1$ choices. The third element can be chosen in $n-2$ ways and so on for the remaining elements. The number of ways of choosing, according to the product rule, must then be

$$N = n * (n - 1) * (n - 2) * \cdots * (n - k + 1) \qquad (4)$$

This product of a sequence of consecutive numbers occurs so often that a special notation is used to represent it. If you multiply all the integers between 1 and n it is called the factorial of n and the notation is:

$$n! = n * (n - 1) * (n - 2) * \cdots * 3 * 2 * 1 \qquad (5)$$

with 0! being defined as equal to 1. As an example the factorial of 5 is $5! = 5 * 4 * 3 * 2 * 1 = 120$. With this notation equation 4 can be written as:

$$N = \frac{n!}{(n - k)!} \qquad (6)$$

The number of ways of selecting all n objects in the set is just $n!$. It may seem like there is only one way to select all the elements in the set but that is only true if the order in which the elements are selected is not important. We have been assuming that order of selection is important so that selecting *dog* and then *cat* from the set of animals is different than selecting *cat* and then *dog*. The number of ways of selecting all the elements in this sense is equal to the number of ways that the elements can be ordered. Each way of ordering the elements is called a permutation. There are $n!$ permutations of n elements, and $n!/(n-k)!$ permutations of n elements taken k at a time. We will give this expression a name, $P(n,k)$, where P stands for permutation, so that

$$P(n,k) = \frac{n!}{(n-k)!} \qquad (7)$$

We should actually refer to the permutations discussed above as linear permutations since they are formed by arranging the elements in some linear order. There is another kind of permutation called a circular permutation where the elements are arranged in a circular order. A common example of this is seating people around a table.

For a circular permutation there is no natural beginning or end of the arrangement and any circular shift of the elements around the circle does not count as a new arrangement. This should mean that there are fewer circular permutations of n elements than there are linear permutations. Indeed there are only $(n-1)!$ circular permutations as opposed to $n!$ linear permutations. To see this, pick a point on the circle to correspond with the start of a linear permutation. If you put one of the linear permutations on the circle then every one of its n possible circular shifts will correspond to a different linear permutation when read from the starting point but each of these shifts is counted as the same circular permutation. For every circular permutation there are n linear permutations therefor the number of circular permutations must be $n!/n = (n-1)!$.

The same argument applies to the number of circular permutations of n elements taken k at a time. There are k circular shifts of the k elements that correspond to different linear permutations but the same circular permutation. So to get the number of circular permutations, divide the linear permutations by k

$$N = \frac{n!}{k(n-k)!} \qquad (8)$$

Combinations

In some cases the order in which k elements are selected from n is not important. Equation 7 counts all the ways that k elements can be selected, including selecting the same elements but in a different order. For example if we are interested in how many 3 letter words can be formed from the English alphabet then the order is important and equation 7 says there are

$$P(26, 3) = \frac{26!}{23!} = 26 * 25 * 24 = 15,600 \qquad (9)$$

possible words. This includes for example the six words: abc, bca, cab, bac, cba, acb. All these words have the same letters. For every set of 3 letters all 6 ways they can be arranged is counted in equation 9. If only the letters are important and not their order then we have to divide by 6 to get $15,600/6 = 2,600$ unique ways of choosing 3 letters.

In general there are $k!$ ways that k elements can be ordered so if the order is not important then the number of ways that k elements can be chosen from n is

$$\frac{n!}{k!(n-k)!} \tag{10}$$

This is equal to the number of k element subsets that can be formed from a set of n elements (the order of elements in a set or subset is irrelevant). It is called the number of combinations of n elements taken k at a time and it is given a special notation:

$$\binom{n}{k} = \frac{n!}{k!(n-k)!} \tag{11}$$

For convenience the equation is also referred to with the function name $C(n, k)$, and stated verbally as "n choose k".

The term $\binom{n}{k}$ is called a binomial coefficient since it appears in the expansion of binomials. If you multiply out the binomial $(x+1)^n$ then the coefficient of x^k in the expansion will be $\binom{n}{k}$. For example $(x+1)^5$ is equal to

$$(x+1)^5 = x^5 + 5x^4 + 10x^3 + 10x^2 + 5x + 1$$

Note that the coefficient of x^3 is $\binom{5}{3} = 10$ which is also the coefficient of x^2 so we have $\binom{5}{3} = \binom{5}{2}$ and in general the binomial coefficients obey:

$$\binom{n}{k} = \binom{n}{n-k} \tag{12}$$

which should be obvious from the way they are defined in equation 11.

Binomial coefficients have many useful and interesting properties. For example summing them for all possible values of k gives

$$\sum_{k=0}^{n} \binom{n}{k} = 2^n \qquad (13)$$

You can see this from the fact that the coefficients appear in the expansion of $(x+1)^n$. Write out the expansion, set $x = 1$ and you get equation 13. An interesting combinatorial interpretation of this equation is covered in one of the problems.

Another useful relationship is

$$\binom{n}{k} = \binom{n-1}{k-1} + \binom{n-1}{k} \qquad (14)$$

This can easily be proven algebraically but there is also a simple combinatorial proof. Let the set of size n be the set of integers from 1 through n. We write such a set of integers as $[n] = \{1, 2, 3, \ldots, n\}$. If we select a k element subset of $[n]$ then it will either contain the integer n or not. If it doesn't contain n then it is equivalent to a k element subset of $[n-1]$ and there are $\binom{n-1}{k}$ of those. If it does contain n then we remove n and we are left with a $k-1$ element subset of $[n-1]$ and there are $\binom{n-1}{k-1}$ of those. This proves the equation.

Equation 14 can be used to easily construct Pascal's triangle shown in figure 1. Row n in the triangle lists the binomial coefficients $\binom{n}{k}$ for $k = 0, 1, \ldots, n$. According to equation 14 the entries in each row should be equal to the sum of the entries to the right and left in the row above. For example entry 1365 in row 15 is equal to the sum of 364 and 1001 in row 14 above it. In this way the triangle can be extended to more rows, keeping in mind that the first and last entry in each row is always 1.

Multinomial Coefficients

The binomial coefficient $\binom{n}{k}$ counts the number of ways you can choose k elements from n when the order of choice does not matter. You can also look at it as the number of ways that a set of n elements can be divided into 2 subsets with k elements in one subset and $n - k$ elements in the other. This of course leads to the question of how many ways you can divide the elements into 3 or more subsets.

Suppose we want to divide the n elements into 3 subsets with k elements in the first subset, l in the second, and $n - k - l$ in the third. The way to do this is straightforward. First select the k elements for the first subset which divides the n elements into a k element subset and a $n - k$ element subset. The number of ways

n																
0								1								
1								1	1							
2							1	2	1							
3						1	3	3	1							
4					1	4	6	4	1							
5				1	5	10	10	5	1							
6			1	6	15	20	15	6	1							
7		1	7	21	35	35	21	7	1							
8	1	8	28	56	70	56	28	8	1							
9	1	9	36	84	126	126	84	36	9	1						
10	1	10	45	120	210	252	210	120	45	10	1					
11	1	11	55	165	330	462	462	330	165	55	11	1				
12	1	12	66	220	495	792	924	792	495	220	66	12	1			
13	1	13	78	286	715	1287	1716	1716	1287	715	286	78	13	1		
14	1	14	91	364	1001	2002	3003	3432	3003	2002	1001	364	91	14	1	
15	1	15	105	455	1365	3003	5005	6435	6435	5005	3003	1365	455	105	15	1

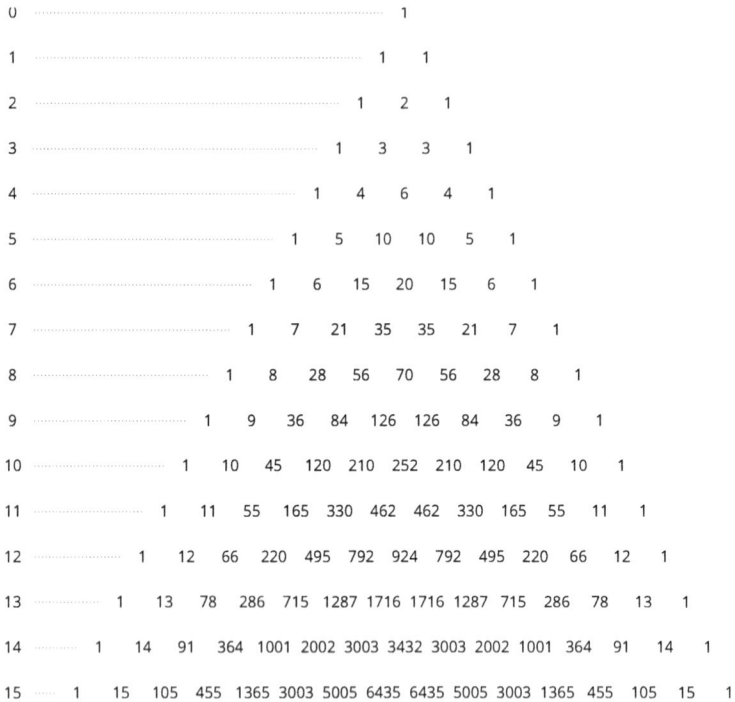

Figure 1: Pascal's triangle.

of doing this is $\binom{n}{k}$. Next select the l elements for the second subset from the $n - k$ element subset. This can be done in $\binom{n-k}{l}$ ways. Now we have 3 subsets of the original n elements of size k, l, and $n - k - l$ and the number of ways of forming these subsets is (using the product rule):

$$\binom{n}{k}\binom{n-k}{l} = \frac{n!}{k!(n-k)!} \cdot \frac{(n-k)!}{l!(n-k-l)!} \tag{15}$$

$$= \frac{n!}{k!l!(n-k-l)!}$$

The extension to more than 3 subsets should be obvious. Let's say we want r subsets with k_i elements in subset $i = 1, 2, \ldots, r$. This should completely divide up the n elements so that $k_1 + k_2 + \cdots + k_r = n$. The number of ways the subsets can be formed is

$$\binom{n}{k_1}\binom{n-k_1}{k_2}\cdots\binom{n-k_1-k_2-\cdots-k_{r-1}}{k_r} \tag{16}$$

When this equation is written out and simplified it becomes

$$\frac{n!}{k_1!k_2!\cdots k_r!} = \binom{n}{k_1, k_2, \cdots, k_r} \tag{17}$$

where we have assigned a new notation on the right side of the equation. This new notation is called a multinomial coefficient. The coefficients appear when you expand a multinomial. In this case they would appear in the expansion of $(x_1 + x_2 + \cdots + x_r)^n$. The coefficient of $x_1^{k_1} x_2^{k_2} \cdots x_r^{k_r}$ in the expansion is given by equation 17.

Multinomial coefficients come up in a variety of contexts. Suppose for example you have a collection of n letters with k_1 of them being a's, k_2 of them being b's, k_3 of them being c's and the remaining $n - k_1 - k_2 - k_3$ all different. How many unique n letter words can you construct? The answer is given by a multinomial coefficient. In this case the n string positions are the objects to choose from. Choose k_1 positions for the a's, k_2 positions for the b's, k_3 positions for the c's and one position for each of the remaining $n - k_1 - k_2 - k_3$ different letters. The number of ways of doing this is

$$\binom{n}{k_1, k_2, k_3, 1, 1, \cdots, 1} = \frac{n!}{k_1! k_2! k_3!} \qquad (18)$$

where the $n - k_1 - k_2 - k_3$ factors equal to $1! = 1$ have not been explicitly included on the right side of the equation. What we have done is divide the set of n string positions into $n - k_1 - k_2 - k_3 + 3$ subsets with k_1 elements in one subset, k_2 in another, k_3 in another, and 1 element in each of $n - k_1 - k_2 - k_3$ subsets.

Balls in Boxes

There are many counting problems that can be formulated in terms of placing balls into boxes (or cells, urns, bins, whatever you like). We will show a systematic classification scheme for these problems in the section on the twelve fold way. In this section we look at some of the more common examples of this sort of problem.

Suppose for example we have 12 people and 12 fruits: 2 mangoes, 4 apples, and 6 oranges. In how many ways can each person get one of the fruits? In this case the people are balls and they can be placed into one of three boxes, a mango box, an apple box, and an orange box. Two people can be put in the mango box, four can go in the apple box and six can go in the orange box. When stated this way you can see that the people are being divided into three groups or sets and the number of ways you can do this is given by a multinomial coefficient (see previous section)

$$\binom{12}{2, 4, 6} = \frac{12!}{2!4!6!} = 13860 \qquad (19)$$

This is an example of placing distinguishable balls into boxes. In general if you have m balls to put into r boxes so that box i has m_i balls then the number of ways to do it is given by the multinomial coefficient

$$\binom{m}{m_1, m_2, \cdots, m_r} = \frac{m!}{m_1!m_2!\cdots m_r!} \qquad (20)$$

Another problem of the same type is finding the number of permutations of a set of letters that may not all be unique. Take the word banana for example. There are three a's, two n's and one b so not all $6! = 720$ permutations of these letters will produce a unique word. You can for example switch the two n's in banana and still get a banana. If the three a's are uniquely identified in some way then they can be arranged in $3! = 6$ ways but if you then remove the identifiers the six arrangements become indistinguishable. So the number of unique permutations must be given by the multinomial coefficient:

$$\binom{6}{3, 2, 1} = \frac{6!}{3!2!1!} = 60 \qquad (21)$$

This kind of permutation of a set of objects where you have groups of objects of the same type is called a multiset permutation. Finding the number of multiset permutations is the same as the fruit distribution problem discussed above and it can also be interpreted in terms of putting balls into boxes. In the banana example the balls are the six string positions and there are three boxes, a b box, an n box and an a box. The b box takes one ball, the n box takes two, and the a box takes three. In general the number of multiset permutations of m objects of r different types, where the number of objects of type i is m_i, is given by the multinomial coefficient in equation 20.

In the two examples discussed above the balls were distinguishable. People are generally distinguishable and letter position in a word is clearly defined and therefor distinguishable. There are other problems where the balls are assumed to be identical or indistinguishable and we have to count the number of ways that m identical balls can be put into r boxes.

In the fruit example suppose we have an unlimited supply of each type of fruit and we only want to know how many ways twelve people can take three kinds of fruit. Again the three fruits are represented by three boxes and the people are now represented by twelve indistinguishable balls. If | symbolizes the side of a box and ∗ symbolizes a ball then one way of distributing 12 balls (people) into 3 boxes (fruits) is as follows: ∗∗|∗∗∗∗∗∗|∗∗∗∗. In this case two people got mangoes (first box), six people got apples (second box), and four people got oranges (third box).

The question is, how how many total ways can the 12 identical balls be put into the 3 boxes? Each distribution can be represented by 14 symbols, two | symbols representing the sides of the boxes and twelve ∗ symbols representing the people. A distribution is fixed by selecting two of the 14 symbols to be | and then letting the rest of the symbols be $*$. The number of ways of selecting two objects from fourteen is

$$\binom{14}{2} = \frac{14!}{2!12!} = \frac{14 * 13}{2} = 91 \qquad (22)$$

This is the total number of ways that 12 people can get three kinds of fruit.

In general if the balls are indistinguishable then every solution of the equation $m_1 + m_2 + \cdots + m_r = m$ where each m_i is an integer, is a way of placing m balls into r boxes. The numbers m_i are called occupancy numbers. When representing this using the stars and bars notation, as in the above example, there will be $r - 1$ vertical bars for the sides of the boxes and m stars for the balls giving a total of $m + r - 1$ symbols. The number of ways of selecting $r - 1$ of the symbols to be bars or m of the symbols to be stars is

$$\binom{m + r - 1}{m} = \binom{m + r - 1}{r - 1} \qquad (23)$$

This is the total number of ways that m indistinguishable balls can be placed into r boxes.

Here is another example that gives a slightly different way of looking at this sort of problem. We have a lottery that involves selecting four numbers from the numbers 1 through 20 where any number can be selected multiple times. In other words let's say the first number selected is 7. The number is then put back into the hopper so that it could be selected again. This is called sampling with replacement.

In most lotteries, the order in which the numbers are selected makes no difference. The only thing that matters is the final set of numbers selected. So this is another example of putting indistinguishable balls into boxes. The boxes in this case are the numbers 1 through 20 and the balls are the four selections. Using equation 23 with $r = 20$ and $m = 4$, the number of possible lottery tickets is

$$\binom{4 + 20 - 1}{4} = \binom{23}{4} = 8855 \qquad (24)$$

Another way of looking at this problem is as follows. Let the four numbers in a lottery drawing be r_1, r_2, r_3, and r_4 where they are labeled so that:

$$1 \leq r_1 \leq r_2 \leq r_3 \leq r_4 \leq 20 \qquad (25)$$

Now define a new set of four numbers: $s_1 = r_1$, $s_2 = r_2 + 1$, $s_3 = r_3 + 2$, $s_4 = r_4 + 3$. These numbers will all be different and they obey the relation:

$$1 \leq s_1 < s_2 < s_3 < s_4 \leq 23 \qquad (26)$$

This shows that there is a one to one correspondence between the act of selecting four numbers from 1 to 20 with replacement and selecting four numbers from 1 to 23 without replacement. The number of ways of selecting four unique numbers from 1 to 23, when the order is not important, is given by the binomial coefficient $\binom{23}{4} = 8855$, which is the same answer we found above.

The general idea is that the number of ways to select m things from r with replacement (repetition allowed) and with order not important, is the same as the number of ways of selecting m things from $m + r - 1$ things with order not important. It is a somewhat different way of looking at equation 23. It is also an example of what is called a bijection in combinatorics. If you can show that there is a one to one mapping between the elements of two sets, so that every element of one set has one and only one corresponding element in the other set, then a counting problem in one set can be translated into a counting problem in the other.

Inclusion Exclusion Principle

If you are told that in a group of dogs there are eight males and twelve females then you know there must be twenty dogs in the group since every dog is either male or female and never both. The male and female set is said to be a partition of the set of dogs. In general however, you can have situations where an object is simultaneously a member of more than one set. Then you can not find the total number of objects by summing the number of objects in each set.

Suppose you are told that in a group of people there are eight who run at least once a week and six who bike at least once a week. Can you conclude from this that there are $8 + 6 = 14$ people who run or bike once a week? No, there is not enough information for such a conclusion. It may be that everyone that bikes also runs in which case there are only 8 people that run or bike once a week, not 14.

What we want to do is extend the sum rule to the case where we have a collection of sets with some elements appearing in more than one set. The simplest example has two sets as shown in figure 2.

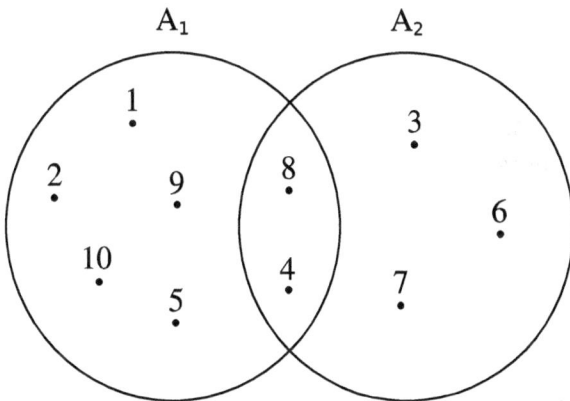

Figure 2: Two set example.

Set A_1 has 7 elements, A_2 has 5 elements and the two sets have elements, 8 and 4, in common. If we sum the number of elements in each set then the elements they have in common will be counted twice. So the number they have in common has to be subtracted. The number of elements in A_1 or A_2 is then $7+5-2=10$.

To get a general formula for two sets, let $|A_i|$ be the number of elements in set A_i then the number of elements in either set A_1 or set A_2 is given by:

$$|A_1 \cup A_2| = |A_1| + |A_2| - |A_1 \cap A_2| \qquad (27)$$

where $|A_1 \cap A_2|$ is the number of elements in both sets A_1 and A_2 i.e. the number of elements in the intersection of the two sets.

To extend this to three sets we need not just the number of elements in two sets at once but also the number of elements in three sets at once. Figure 3 shows an example.

It is similar to the two set example but now there is a third set A_3 with 5 elements. A_3 has two elements in common with both A_1 and A_2 and there is one element that is common to all three sets. If you sum the elements in all three sets then you will count elements 5, 7, and 8 twice and element 4 three times. Subtracting intersections between pairs of sets will zero out element 4 completely since it appears in three intersections. Element 4 is added back by adding the number of elements in the intersection of all three sets.

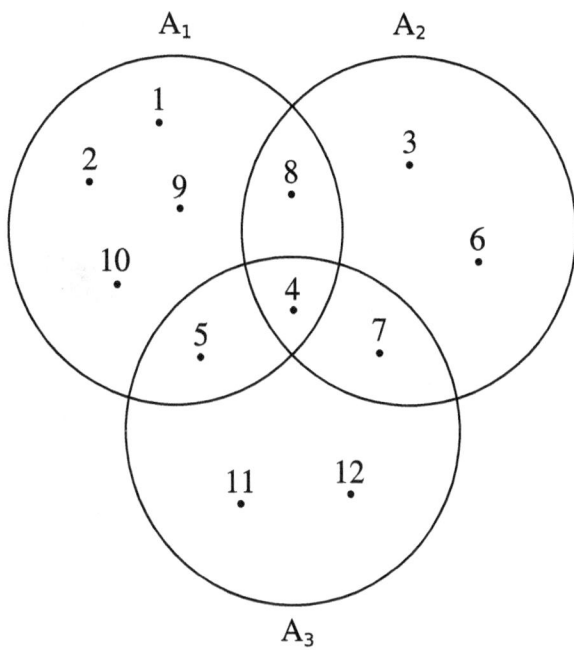

Figure 3: Three set example.

The general formula for three sets is:

$$|A_1 \cup A_2 \cup A_3| = |A_1| + |A_2| + |A_3| - |A_1 \cap A_2| - |A_1 \cap A_3|$$

$$- |A_2 \cap A_3| + |A_1 \cap A_2 \cap A_3| \qquad (28)$$

In general, to count the number of elements in n sets we have the following formula:

$$|A_1 \cup A_2 \cup \cdots A_n| = \sum_{i=1}^{n} (-1)^{i-1} \sum_{j_1, j_2, \cdots, j_i} |A_{j_1} \cap A_{j_2} \cap \cdots A_{j_i}|$$

$$(29)$$

where the inner summation is over all i element subsets of $(1, 2, \ldots, n)$. To prove this formula you only have to show that any element in the union contributes a 1 to the summation on the right hand side. We will sketch the proof but it can easily be skipped without loss of continuity or increase in confusion.

Take an element x of the union that is a member of s of the sets. The $|A_i|$ terms will contribute a $\binom{s}{1} = s$ factor when counting the element. There will be a $\binom{s}{2}$ factor from the $|A_i \cap A_j|$ terms, a $\binom{s}{3}$ factor from the $|A_i \cap A_j \cap A_k|$ terms and so on. The count for x is then equal to:

$$\binom{s}{1} - \binom{s}{2} + \binom{s}{3} - \cdots - (-1)^s \binom{s}{s} \qquad (30)$$

You can see that this equation is always equal to 1 by comparing it to the expansion of $(y - 1)^s$.

$$(y - 1)^s = y^s - \binom{s}{1} y^{s-1} + \binom{s}{2} y^{s-2} - \binom{s}{3} y^{s-3} + \cdots$$

$$+ (-1)^s \binom{s}{s} \tag{31}$$

If you subtract this equation from 1 and set $y = 1$ you get equation 30. This means equation 30 is equal to $1 - (1 - 1)^s = 1$ and it completes the proof.

An interesting application of equation 29 is counting the number of derangements. A derangement is a permutation of the numbers $1, 2, \ldots, n$ where no number equals its order in the permutation, i.e. none of the numbers stay in the same position. When $n = 2$ the only derangement is $2, 1$. When $n = 3$ there are two derangements: $2, 3, 1$ and $3, 1, 2$. So for general n how many derangements are there? First let's calculate how many permutations leave at least one number in a fixed position. Let A_i be the set of permutations that leave the number i fixed. With i fixed, the other numbers can be permuted in $(n - 1)!$ ways so this is the size of A_i. The sum of the sizes of all the A_i is then equal to $n(n - 1)! = n!$. The set of permutations that leave both i and j fixed is $A_i \cap A_j$. With i and j fixed the other numbers can be permuted in $(n - 2)!$ ways so this is the size of $A_i \cap A_j$. There are $\binom{n}{2}$ ways to fix a pair of numbers so summing them all up gives

$$\binom{n}{2} (n - 2)! = \frac{n!}{2!} \tag{32}$$

Likewise the contribution from three fixed numbers is

$$\binom{n}{3}(n-3)! = \frac{n!}{3!} \tag{33}$$

and so on. Putting all this into equation 29 gives the number of permutations with at least one fixed number.

$$n! \left[\frac{1}{1!} - \frac{1}{2!} + \frac{1}{3!} - (-1)^n \frac{1}{n!} \right] \tag{34}$$

The number of derangements is the total number of permutations, $n!$, minus this number. Call D_n the number of derangements then

$$D_n = n! \sum_{i=0}^{n} \frac{(-1)^i}{i!} \tag{35}$$

To find the probability that a random permutation is a derangement divide both sides of the equation by $n!$

$$\frac{D_n}{n!} = \sum_{i=0}^{n} \frac{(-1)^i}{i!} \tag{36}$$

The sum on the right hand side of this equation converges to $e^{-1} = .367879$ as n goes to infinity. This means that for large n the probability of a random permutation being a derangement is about 37%.

Pigeon Hole Principle

The pigeon hole principle is probably the easiest concept to understand in all of combinatorics. In its simplest form it says that if you have a bunch of objects that you want to put in boxes and there are fewer boxes than objects then at least one of the boxes will have to hold multiple objects. More formally we can say that if there are n boxes and $k > n$ objects then at least one of the boxes must contain 2 or more objects. We can generalize this a bit by saying that if we have n boxes and more than kn objects then at least one of the boxes must contain $k + 1$ or more objects.

Another form of the pigeon hole principle can be illustrated by the following example. Suppose we have a drawer with three pairs of gloves. If gloves are removed from the drawer at random, what is the minimum number that have to be removed to ensure a matching pair? In the worst case the first three removed will all be of the same type. The next one will then have to be of the opposite type and thus form a matching pair.

To generalize this example suppose we have a multiset of k different kinds of things with n_i things of type $i = 1, 2, \ldots, k$. What is the minimum size subset that ensures at least two of the same kind of thing? There are only k different things so a set of size $k + 1$ must contain two of a kind. What is the minimum size set to ensure three of a kind? The largest possible set with only two of a kind is $2k$. A set of size $2k + 1$ must contain at least one three of a kind. What is the minimum size set to ensure at least two different kinds of things? If n_i is the largest number of things of the same type then a set of size $n_i + 1$ will ensure at least two different kinds of things. The way to generalize this further should be clear at this point.

Stirling Numbers of the Second Kind

How many ways can you partition a set of size n into k nonempty subsets? The question is equivalent to asking for the number of ways of putting n distinct balls into k identical urns so that each urn has at least one ball. The answer is given by the Stirling number of the second kind $S(n, k)$. We will derive a formula for $S(n, k)$ by using the inclusion exclusion principle. To begin, the number of ways to put n distinct balls into k distinct urns without restriction is k^n. The number of ways to put the balls into the urns so that at least i of the urns remain empty is $(k - i)^n$. There are $\binom{k}{i}$ ways that i of the urns can remain empty. So from the inclusion exclusion principle the number of ways to distribute the balls so that at least one urn remains empty is

$$\sum_{i=1}^{k-1}(-1)^{i-1}\binom{k}{i}(k-i)^n \qquad (37)$$

Now subtract this equation from k^n to get the number of ways to distribute n distinct balls into k distinct urns with at least one ball in each urn. The answer is

$$\sum_{i=0}^{k-1}(-1)^i\binom{k}{i}(k-i)^n \qquad (38)$$

This equation should equal $k!S(n,k)$ since in the case of the Stirling number the distribution is into urns that are identical. The equation for $S(n,k)$ must then be

$$S(n,k) = \frac{1}{k!}\sum_{i=0}^{k-1}(-1)^i\binom{k}{i}(k-i)^n \qquad (39)$$

An equivalent form of this equation is

$$S(n,k) = \frac{1}{k!}\sum_{i=1}^{k}(-1)^{k-i}\binom{k}{i}i^n \qquad (40)$$

The following formulas are for the first four k values.

$$S(n,1) = 1 \qquad (41)$$

$$S(n,2) = \frac{1}{2!}(2^n - 2) = 2^{n-1} - 1$$

$$S(n,3) = \frac{1}{3!}(3^n - 3\cdot 2^n + 3)$$

$$S(n,4) = \frac{1}{4!}(4^n - 4\cdot 3^n + 6\cdot 2^n - 4)$$

n	2	3	4	5	k 6	7	8	9
2	1							
3	3	1						
4	7	6	1					
5	15	25	10	1				
6	31	90	65	15	1			
7	63	301	350	140	21	1		
8	127	966	1701	1050	266	28	1	
9	255	3025	7770	6951	2646	462	36	1
10	511	9330	34105	42525	22827	5880	750	45

Table 1: Stirling numbers of the second kind, $S(n, k)$

The numbers can also be calculated using the following recurrence

$$S(n, k) = S(n - 1, k - 1) + kS(n - 1, k) \qquad (42)$$

with the initial conditions, $S(n, 0) = 0$, $S(n, 1) = 1$, and $S(n, n) = 1$ for all n. Table 1 lists the values of $S(n, k)$ for $n = 2, 3, \ldots, 10$.

The recurrence in eq. 42 has a simple combinatorial explanation. Note first of all that $S(n, k)$ can be equivalently defined to be the number of ways that the set $[n] = \{1, 2, 3, \ldots, n\}$ can be partitioned into k nonempty sets. In any such partition the number n will either be in a set by itself or not. If it is, then removing that set will leave a partition of $[n-1]$ into $k-1$ sets and the number of such partitions is $S(n-1, k-1)$, which is the first term on the right hand side of eq. 42.

If n is not in a set by itself and we remove it from the set it is in, then we are left with a partition of $[n-1]$ into k sets and there are $S(n-1,k)$ of those. But n can be in any one of those k sets so the number of ways to get an $S(n,k)$ partition from a $S(n-1,k)$ partition is $kS(n-1,k)$ which is the second term on the right hand side of eq. 42. This establishes the equality.

The number of ways to distribute n distinct balls into k distinct boxes with no empty boxes is given by:

$$S(n,k)k! \qquad (43)$$

To see this, note that for every $S(n,k)$ distribution we can label the k identical boxes in $k!$ ways so the number of distributions into distinct boxes must be $S(n,k)k!$.

The number of ways to distribute n distinct balls into k distinct boxes with no restrictions, meaning some boxes can remain empty, is k^n. Using the result in eq. 43 we can also express k^n in terms of the $S(n,k)$ numbers.

Note first of all that any such distribution can be categorized in terms of how many boxes contain balls. If exactly j boxes contain balls ($k-j$ boxes remain empty) those j boxes can be chosen in $\binom{k}{j}$ ways and the balls can be distributed into the boxes in $S(n,j)j!$ ways. So the number of distributions with j boxes containing balls is $S(n,j)j!\binom{k}{j}$. Summing over all possible values of j we have

$$k^n = \sum_{j=1}^{k} S(n,j) j! \binom{k}{j} \qquad (44)$$

This result is used in one of the more challenging problems in the problems section.

Bell Numbers

If you sum $S(n,k)$ over all possible values of k you get the total number of ways a set of n elements can be partitioned into nonempty sets. This number is called the Bell number, $B(n)$

$$B(n) = \sum_{k=1}^{n} S(n,k) \qquad (45)$$

Starting with $B(0) = 1$ the Bell numbers can be calculated using the following recurrence.

$$B(n) = \sum_{k=0}^{n-1} \binom{n-1}{k} B(k) \qquad (46)$$

The values of the first few Bell numbers are shown in table 2.

Here for example are all the possible partitions of the set $\{1, 2, 3, 4\}$ into nonempty sets. There are $B(4) = 15$.

n	1	2	3	4	5	6	7	8	9
B(n)	1	2	5	15	52	203	877	4140	21147

Table 2: Bell numbers, $B(n)$

1. $\{1\}\ \{2\}\ \{3\}\ \{4\}$

2. $\{1\}\ \{2\}\ \{3,4\}$

3. $\{1\}\ \{3\}\ \{2,4\}$

4. $\{1\}\ \{4\}\ \{2,3\}$

5. $\{2\}\ \{3\}\ \{1,4\}$

6. $\{2\}\ \{4\}\ \{1,3\}$

7. $\{3\}\ \{4\}\ \{1,2\}$

8. $\{1\}\ \{2,3,4\}$

9. $\{2\}\ \{1,3,4\}$

10. $\{3\}\ \{1,2,4\}$

11. $\{4\}\ \{1,2,3\}$

12. $\{1,2\}\ \{3,4\}$

13. $\{1,3\}\ \{2,4\}$

14. $\{1,4\}\ \{2,3\}$

15. $\{1,2,3,4\}$

From the above list we can see that there is one partition into one set, corresponding to $S(4, 1) = 1$. There are seven partitions into two sets corresponding to $S(4, 2) = 7$. There are six partitions into three sets corresponding to $S(4, 3) = 6$ and one partition into four sets corresponding to $S(4, 4) = 1$. So we have $B(4) = 1 + 7 + 6 + 1 = 15$.

Integer Partitions

A partition of the integer n is a set of one or more nonzero integers that sum to n. For example the integer 4 has the following five partitions: $\{1, 1, 1, 1\}$, $\{1, 1, 2\}$, $\{1, 3\}$, $\{2, 2\}$, $\{4\}$. The partitions are really multisets since an integer may appear more than once in the set. Like any set the order of the elements makes no difference. The integers (elements) in the set are called the parts of the partition.

Another way to look at the number of partitions of n is that it is equal to the number of ways to partition a set of size n when the elements of the set are identical or the identity of the elements is not important. For example in the set partitions of [4] on page 42 we see that when the identity of the elements is not important then we only have the following unique partitions:

1. $\{*\}\,\{*\}\,\{*\}\,\{*\}$

2. $\{*\}$ $\{*\}$ $\{**\}$

3. $\{*\}$ $\{***\}$

4. $\{**\}$ $\{**\}$

5. $\{****\}$

Replace each set with the number of elements in the set and you get the partitions of 4 that we listed above. In the framework of putting balls into boxes, the number of integer partitions of n is the number of ways to put n identical balls into identical boxes. You can put them all into one box, each into a separate box or anything in between. The following is a list of partitions.

Partitions for $n = 1, 2, \ldots, 10$.

1. $\{1\}$

2. $\{1,1\}$, $\{2\}$

3. $\{1,1,1\}$, $\{1,2\}$, $\{3\}$

4. $\{1,1,1,1\}$, $\{1,1,2\}$, $\{1,3\}$, $\{2,2\}$, $\{4\}$

5. $\{1,1,1,1,1\}$, $\{1,1,1,2\}$, $\{1,1,3\}$, $\{1,2,2\}$, $\{1,4\}$, $\{2,3\}$, $\{5\}$

6. $\{1,1,1,1,1,1\}$, $\{1,1,1,1,2\}$, $\{1,1,1,3\}$, $\{1,1,2,2\}$, $\{1,1,4\}$, $\{1,2,3\}$, $\{2,2,2\}$, $\{1,5\}$, $\{2,4\}$, $\{3,3\}$, $\{6\}$

7. $\{1,1,1,1,1,1,1\}$, $\{1,1,1,1,1,2\}$, $\{1,1,1,1,3\}$, $\{1,1,1,2,2\}$, $\{1,1,1,4\}$, $\{1,1,2,3\}$, $\{1,2,2,2\}$, $\{1,1,5\}$, $\{1,2,4\}$, $\{1,3,3\}$, $\{2,2,3\}$, $\{1,6\}$, $\{2,5\}$, $\{3,4\}$, $\{7\}$

8. $\{1,1,1,1,1,1,1,1\}$, $\{1,1,1,1,1,1,2\}$, $\{1,1,1,1,1,3\}$,
   $\{1,1,1,1,2,2\}$, $\{1,1,1,1,4\}$, $\{1,1,1,2,3\}$, $\{1,1,2,2,2\}$,
   $\{1,1,1,5\}$, $\{1,1,2,4\}$, $\{1,1,3,3\}$, $\{1,2,2,3\}$, $\{2,2,2,2\}$,
   $\{1,1,6\}$, $\{1,2,5\}$, $\{1,3,4\}$, $\{2,2,4\}$, $\{2,3,3\}$, $\{1,7\}$, $\{2,6\}$,
   $\{3,5\}$, $\{4,4\}$, $\{8\}$

9. $\{1,1,1,1,1,1,1,1,1\}$, $\{1,1,1,1,1,1,1,2\}$, $\{1,1,1,1,1,1,3\}$,
   $\{1,1,1,1,1,2,2\}$, $\{1,1,1,1,1,4\}$, $\{1,1,1,1,2,3\}$, $\{1,1,1,2,2,2\}$,
   $\{1,1,1,1,5\}$, $\{1,1,1,2,4\}$, $\{1,1,1,3,3\}$, $\{1,1,2,2,3\}$, $\{1,2,2,2,2\}$,
   $\{1,1,1,6\}$, $\{1,1,2,5\}$, $\{1,1,3,4\}$, $\{1,2,2,4\}$, $\{1,2,3,3\}$,
   $\{2,2,2,3\}$, $\{1,1,7\}$, $\{1,2,6\}$, $\{1,3,5\}$, $\{1,4,4\}$, $\{3,3,3\}$,
   $\{2,2,5\}$, $\{2,3,4\}$, $\{1,8\}$, $\{2,7\}$, $\{3,6\}$, $\{4,5\}$, $\{9\}$

10. $\{1,1,1,1,1,1,1,1,1,1\}$, $\{1,1,1,1,1,1,1,1,2\}$, $\{1,1,1,1,1,1,1,3\}$,
    $\{1,1,1,1,1,1,2,2\}$, $\{1,1,1,1,1,1,4\}$, $\{1,1,1,1,1,2,3\}$,
    $\{1,1,1,1,2,2,2\}$, $\{1,1,1,1,1,5\}$, $\{1,1,1,1,2,4\}$, $\{1,1,1,1,3,3\}$,
    $\{1,1,1,2,2,3\}$, $\{1,1,2,2,2,2\}$, $\{1,1,1,1,6\}$, $\{1,1,1,2,5\}$,
    $\{1,1,1,3,4\}$, $\{1,1,2,2,4\}$, $\{1,1,2,3,3\}$, $\{1,2,2,2,3\}$, $\{2,2,2,2,2\}$,
    $\{1,1,1,7\}$, $\{1,1,2,6\}$, $\{1,1,3,5\}$, $\{1,1,4,4\}$, $\{1,2,2,5\}$,
    $\{1,2,3,4\}$, $\{1,3,3,3\}$, $\{2,2,2,4\}$, $\{2,2,3,3\}$, $\{1,1,8\}$, $\{1,2,7\}$,
    $\{1,3,6\}$, $\{1,4,5\}$, $\{2,2,6\}$, $\{2,3,5\}$, $\{2,4,4\}$, $\{3,3,4\}$,
    $\{1,9\}$, $\{2,8\}$, $\{3,7\}$, $\{4,6\}$, $\{5,5\}$, $\{10\}$

We will use the notation $p(n)$ to denote the number of
partitions of n. Count the partitions of 10 in the above
list and you will find that $p(10) = 42$. Table 3 shows
the value of $p(n)$ for $n = 1, 2, \ldots, 21$.

n	1	2	3	4	5	6	7	8	9	10	11	12
p(n)	1	2	3	5	7	11	15	22	30	42	56	77

n	13	14	15	16	17	18	19	20	21
p(n)	101	135	176	231	297	385	490	627	792

Table 3: Integer partition numbers, $p(n)$

Now you're probably wondering what the formula for $p(n)$ looks like but unfortunately there is no formula. The closest thing to a formula is the following asymptotic formula

$$p(n) \sim \frac{1}{4n\sqrt{3}} \exp\left(\pi\sqrt{\frac{2n}{3}}\right) \qquad (47)$$

But this formula is only good for getting estimates of $p(n)$ for large values of n. The formula is more useful in number theory than in combinatorics.

It is possible to find the values of $p(n)$ using what are called generating functions. This is a vast subject that could fill a book by itself. See some of the references for more on the subject. We will only touch on generating functions briefly in some of the problems.

There are also some recurrence equations that can be used to calculate $p(n)$ in terms of other values of $p(n)$ such as $p(6) = p(5) + p(4) - p(1) = 7 + 5 - 1 = 11$. The interested reader should consult the book on integer partitions by George Andrews listed in the reference section.

Oftentimes it is more useful to know $p(n, k)$, the number of partitions of n into k parts. Looking at the list of partitions of 10 shown above, we see that $p(10, 1) = 1$, $p(10, 2) = 5$, $p(10, 3) = 8$, $p(10, 4) = 9$, $p(10, 5) = 7$, $p(10, 6) = 5$, $p(10, 7) = 3$, $p(10, 8) = 2$, $p(10, 9) = 1$, $p(10, 10) = 1$. This exhausts all the possibilities and if we sum them all up we get $1 + 5 + 8 + 9 + 7 + 5 + 3 + 2 + 1 + 1 = 42 = p(10)$. In general we have

$$p(n) = \sum_{k=1}^{n} p(n, k) \qquad (48)$$

As with $p(n)$, there are no formulas for $p(n, k)$ for arbitrary values of n and k but there are some formulas. You should, for example, be able to convince yourself that $p(n, n) = p(n, n-1) = p(n, 1) = 1$. Another fairly straightforward result is

$$p(n, 2) = \begin{cases} n/2 & n = \text{even} \\ (n - 1)/2 & n = \text{odd} \end{cases} \qquad (49)$$

It becomes increasingly more difficult to find formulas for $k > 2$. For $k = 3$ we have $p(n, 3) = \lfloor n^2/12 \rceil$ where $\lfloor x \rceil$ means take the nearest integer to x.

There are two useful recursion equations for calculating $p(n, k)$. The first one is:

$$p(n, k) = p(n - 1, k - 1) + p(n - k, k) \qquad (50)$$

The justification for this equation is as follows. Every partition of n into k parts will either have some parts equal to 1 or all parts greater than 1. If it has parts equal to 1 then delete one of those parts and you are left with a partition of $n - 1$ into $k - 1$ parts. If all parts are greater than 1 then subtract 1 from each part

n	k										
	2	3	4	5	6	7	8	9	10	11	12
2	1										
3	1	1									
4	2	1	1								
5	2	2	1	1							
6	3	3	2	1	1						
7	3	4	3	2	1	1					
8	4	5	5	3	2	1	1				
9	4	7	6	5	3	2	1	1			
10	5	8	9	7	5	3	2	1	1		
11	5	10	11	10	7	5	3	2	1	1	
12	6	12	15	13	11	7	5	3	2	1	1

Table 4: Integer partition numbers, $p(n,k)$

and you are left with a partition of $n - k$ into k parts. The second recursion equation is

$$p(n,k) = \sum_{j=1}^{k} p(n-k, j) \qquad (51)$$

The justification for this equation is as follows. For every partition of n into k parts subtract 1 from each of the parts and discard any resulting 0's. The resulting partitions are partitions of $n - k$ into at most k parts. Table 4 shows the $p(n,k)$ values for n up to 12. The table can be extended by using one of the above two recursion equations.

The Twelve Fold Way

The concept of putting balls into boxes covers so many combinatorics problems that a twelve fold classification scheme for these problems has been developed. To see how the classification scheme works note first of all that both balls and boxes can be distinct or indistinct (identical). Let $\textcircled{D}$ and $\textcircled{I}$ represent distinct and indistinct balls and let $\boxed{D}$ and $\boxed{I}$ represent distinct and indistinct boxes then we have the following four possible classifications for a balls into boxes problem.

$\textcircled{D}$ $\boxed{D}$ Both balls and boxes are distinct.

$\textcircled{D}$ $\boxed{I}$ Balls are distinct and boxes are indistinct.

$\textcircled{I}$ $\boxed{D}$ Balls are indistinct and boxes are distinct.

$\textcircled{I}$ $\boxed{I}$ Both balls and boxes are indistinct.

We can also classify a problem in terms of the number of balls that are allowed in a box. Let x be the number of balls per box then we have the following three additional classifications.

1. **General**: $x = 0, 1, 2, \ldots$
   Any number of balls can go into a box. Some boxes can remain empty.

2. **One to One:** $x = 0, 1$
   Every box contains at most one ball. Some boxes can remain empty.

3. **Onto:** $x = 1, 2, 3, \ldots$
   Every box must contain at least one ball. No box can remain empty.

The four types of ball and box classifications together with the three classifications in terms of the number of balls per box create the twelve fold classification scheme shown in figure 4. The formulas in the figure are defined in the following sections.

This twelve fold classification scheme is usually referred to as the twelve fold way. The scheme was initially proposed by Gian-Carlo Rota. According to Richard Stanley (see Richard P. Stanley, Enumerative Combinatorics, Vol 1, 2nd edition, p. 79) the term twelve fold way was suggested by Joel Spencer. There have been various proposals to extend the classification scheme. Robert A. Proctor has extended it to the thirtyfold way (see Proctor, Robert A. "Let's Expand Rota's Twelve Fold Way For Counting Partitions!." arXiv preprint math/0606404 (2006)). We will add a few more classifications to the twelve fold way in the list below. These will suffice for most of the problems we will discuss in this book.

Probably the best way to learn the formulas and their applicability in the seventeen fold way below is not to

Balls

		Distinct	Identical	Balls/Box
Boxes	**Distinct**	n^k	$\dbinom{n+k-1}{k-1}$	General: $0, 1, 2, \ldots$
		$\dfrac{n!}{(n-k)!}$	$\dbinom{n}{k}$	One to One: $0, 1$
		$S(n,k)k!$	$\dbinom{n-1}{k-1}$	Onto: $1, 2, 3, \ldots$
	Identical	$\displaystyle\sum_{i=1}^{k} S(n,i)$	$\displaystyle\sum_{i=1}^{k} p(n,i)$	General: $0, 1, 2, \ldots$
		$\begin{array}{ll} 1 & n \le k \\ 0 & n > k \end{array}$	$\begin{array}{ll} 1 & n \le k \\ 0 & n > k \end{array}$	One to One: $0, 1$
		$S(n,k)$	$p(n,k)$	Onto: $1, 2, 3, \ldots$

Figure 4: The Twelve Fold Way. Formulas are explained in the following sections.

intentionally try to memorize them, but to learn them in the process of working problems. When starting on a new problem, think of the essential model that the problem represents, then find the matching question, and that will give you the formula. As you work more problems, you'll remember more.

1. k Distinct Balls → n Distinct Boxes
General: 0,1,2,... Balls/Box

$$n^k \qquad (52)$$

- How many ways can you place k distinct balls into n distinct boxes with no restrictions?

- How many k letter words can you make with n types of letters?

- How many functions are there from $[k]$ to $[n]$.

2. n Indistinct Balls → k Distinct Boxes
General: 0,1,2,... Balls/Box

$$\binom{n+k-1}{k-1} = \frac{(n+k-1)!}{n!(k-1)!} \qquad (53)$$

- How many ways can you place n indistinct balls into k distinct boxes with no restrictions?

- How many multisets of size n can you create by sampling with replacement from a set of size k?

- How many compositions of n into k parts are there?

3. k Distinct Balls $\rightarrow$ n Distinct Boxes
One to One: 0,1 Balls/Box

$$\frac{n!}{(n-k)!} \qquad (54)$$

- How many ways can you place k distinct balls into $n \geq k$ distinct boxes with at most one ball per box?

- How many lists of length k can you construct from a set of size n?

- How many permutations are there of n distinct things taken k at a time?

- How many words of length k can you construct from an alphabet of size n with no symbol used more than once?

4. k Indistinct Balls → n Distinct Boxes One to One: 0,1 Balls/Box

$$\binom{n}{k} = \frac{n!}{k!(n-k)!} \tag{55}$$

- How many ways can you place k indistinct balls into $n \geq k$ distinct boxes with no more than one ball in each box?

- How many subsets of size k can you produce from a set of size n?

5. n Distinct Balls → k Distinct Boxes Onto: 1,2,3,... Balls/Box

$$S(n,k)k! \tag{56}$$

$S(n,k)$ is a Stirling number of the second kind. It is the number of ways to partition a set of size n into k nonempty subsets.

- How many ways can you place n distinct balls into k distinct boxes with at least one ball in each box?

- How many ways can you create k lists from a set of size n with each list being greater than or equal to one in length?

6. n Indistinct Balls → k Distinct Boxes
Onto: 1,2,3,... Balls/Box

$$\binom{n-1}{k-1} = \frac{(n-1)!}{(k-1)!(n-k)!} \qquad (57)$$

- How many ways can you place n indistinct balls into k distinct boxes with at least one ball in each box?

- How many compositions of n into k nonzero parts are there?

7. n Distinct Balls → k Indistinct Boxes
General: 0,1,2,... Balls/Box

$$\sum_{i=1}^{k} S(n, i) \qquad (58)$$

$S(n, i)$ is a Stirling number of the second kind. It is the number of ways to partition a set of size n into i nonempty subsets.

- How many ways can you place n distinct balls into k indistinct boxes with no restrictions?

8. n Indistinct Balls → k Indistinct Boxes General: 0,1,2,... Balls/Box

$$\sum_{i=1}^{k} p(n,i) \tag{59}$$

$p(n,i)$ is the number of ways to partition the integer n into i parts.

- How many ways can you place n indistinct balls into k indistinct boxes with no restrictions?

9. n Distinct Balls → k Indistinct Boxes One to One: 0,1 Balls/Box

$$\begin{array}{ll} 1 & \text{if } n \le k \\ 0 & \text{if } n > k \end{array} \tag{60}$$

- How many ways can you place n distinct balls into k indistinct boxes with no more than one ball in each box?

10. n Indistinct Balls → k Indistinct Boxes
One to One: 0,1 Balls/Box

$$1 \quad \text{if } n \leq k \qquad (61)$$
$$0 \quad \text{if } n > k$$

- How many ways can you place n indistinct balls into k indistinct boxes with no more than one ball in each box?

11. n Distinct Balls → k Indistinct Boxes
Onto: 1,2,3,... Balls/Box

$$S(n, k) \qquad (62)$$

$S(n, k)$ is a Stirling number of the second kind. It is the number of ways to partition a set of size n into k nonempty subsets.

- How many ways can you place n distinct balls into k indistinct boxes with at least one ball in each box?

- How many ways can you partition a set of size n into k nonempty subsets?

12. n Indistinct Balls → k Indistinct Boxes
Onto: 1,2,3,... Balls/Box

$$p(n, k) \tag{63}$$

$p(n, k)$ is the number of ways to partition the integer n into k parts.

- How many ways can you place n indistinct balls into k indistinct boxes with at least one ball in each box?

- How many ways can you partition the integer n into k parts?

13. n Distinct Balls → k Distinct Boxes
a_i balls in box i

$$\frac{n!}{a_1! a_2! \cdots a_k!} \tag{64}$$

- How many ways can you place n distinct balls into k distinct boxes so that bin i contains a_i balls with $a_1 + a_2 + \cdots + a_k = n$?

- How many ways can you order a multiset of n objects of k different types with a_i objects of type i?

- How many words of length n can you construct using a multi-alphabet with k kinds of letters and a_i copies of letter i?

- How many ways can you split a set of n objects into k subsets with the sizes of the subsets given by a_i.

14. Circular permutations of n things taken k at a time

$$\frac{n!}{k(n-k)!} \qquad (65)$$

- How many circular lists of size k can you construct from a set of $n \geq k$ distinct things?

- How many different bracelets can you make using k beads from a set of $n \geq k$ distinct beads?

15. Total number of subsets of a set of size n

$$\sum_{k=0}^{n} \binom{n}{k} = 2^n \qquad (66)$$

- How many subsets can you create from a set of size n?

- How many words of length n can you make using only two kinds of letters?

- How many n digit binary numbers are there?

16. Number of ways to partition a set

$$B(n) = \sum_{k=1}^{n} S(n, k) \qquad (67)$$

$B(n)$ is called a Bell number. It is the number of ways to partition a set of size n into nonempty subsets.

- How many ways can a set of size n be partitioned into nonempty subsets.

17. Number of partitions of an integer

$$p(n) = \sum_{k=1}^{n} p(n, k) \tag{68}$$

$p(n)$ is the number of ways to partition the integer n.

- How many ways can the integer n be partitioned into nonzero parts.

Problem 1. If there are 6 routes from Yellowstone National Park to Glacier National Park, and 4 routes from Glacier National Park to Banff National Park, how many ways can you go from Yellowstone to Banff via Glacier?

Answer. By the product rule, there are $6 \cdot 4 = 24$ ways.

Problem 2. If a nickel and a dime are tossed, how many ways can the coins fall?

Answer. Both the nickel and the dime can fall with heads up or tails up. So we are choosing from 2 sets each of size 2. The number of ways is therefore $2 \cdot 2 = 4$. If heads is H and tails is T, then all the ways is shown in table 5.

Nickel	Dime
H	H
H	T
T	H
T	T

Table 5: Ways for Problem 2.

Problem 3. How many ways can we choose a consonant and a vowel from letters of the word *almost*?

Answer. There are 2 vowels and 4 consonants, so we can choose a consonant and a vowel in $2 \cdot 4 = 8$ ways.

Problem 4. How many ways can we choose a consonant and a vowel from letters of the word *orange*?

Answer. There are 3 vowels and 3 consonants, so we can choose a consonant and a vowel in $3 \cdot 3 = 9$ ways.

Problem 5. If 2 dice are thrown together, in how many ways can they fall?

Answer. The first die can fall in 6 ways, and the second in 6 ways, so by the product rule they can fall together in $6 \cdot 6 = 36$ ways. All the ways are shown below.

$$
\begin{array}{cccccc}
\{1,1\} & \{1,2\} & \{1,3\} & \{1,4\} & \{1,5\} & \{1,6\} \\
\{2,1\} & \{2,2\} & \{2,3\} & \{2,4\} & \{2,5\} & \{2,6\} \\
\{3,1\} & \{3,2\} & \{3,3\} & \{3,4\} & \{3,5\} & \{3,6\} \\
\{4,1\} & \{4,2\} & \{4,3\} & \{4,4\} & \{4,5\} & \{4,6\} \\
\{5,1\} & \{5,2\} & \{5,3\} & \{5,4\} & \{5,5\} & \{5,6\} \\
\{6,1\} & \{6,2\} & \{6,3\} & \{6,4\} & \{6,5\} & \{6,6\}
\end{array}
$$

Problem 6. How can you represent the 31 days of a month using the faces of 2 cubes? (This problem was inspired by **The Cuboid Calendar** puzzle in *The Great Sherlock Holmes Puzzle Book* by Dr Gareth Moore, pg 30)

Answer. By the previous problem we know that a pair of dice can fall together in $6 \cdot 6 = 36$ ways. To encode the numbers $1, 2, \ldots 31$ on the faces of 2 cubes we can use the cubes to represent a 2 digit senary (base 6) number. This is easily done by labeling the 6 faces of each cube with the numbers $0, 1, 2, 3, 4, 5$ so one cube will provide the 6^1 digit, while the other, the 6^0 digit. A listing of the senary to decimal conversion is below.

$$01 \to 1, \quad 02 \to 2, \quad 03 \to 3, \quad 04 \to 4, \quad 05 \to 5,$$
$$10 \to 6, \quad 11 \to 7, \quad 12 \to 8, \quad 13 \to 9, \quad 14 \to 10,$$
$$15 \to 11, \quad 20 \to 12, \quad 21 \to 13, \quad 22 \to 14, \quad 23 \to 15,$$
$$24 \to 16, \quad 25 \to 17, \quad 30 \to 18, \quad 31 \to 19, \quad 32 \to 20,$$
$$33 \to 21, \quad 34 \to 22, \quad 35 \to 23, \quad 40 \to 24, \quad 41 \to 25,$$
$$42 \to 26, \quad 43 \to 27, \quad 44 \to 28, \quad 45 \to 29, \quad 50 \to 30,$$
$$51 \to 31$$

Problem 7. A die of 6 faces and a teetotum of 8 faces are thrown. In how many ways can they fall?

Answer. $6 \cdot 8 = 48$ ways.

Problem 8. There are 3 major routes to the top of Longs Peak: the Keyhole route, the Loft route, and Keplinger's Couloir. In how many ways can a person go up and down?

Answer. $3 \cdot 3 = 9$ ways.

Problem 9. In how many ways can 2 prizes be given to a class of 10 kids, without giving both to the same kid?

Answer. For the first prize we have a set of 10 kids to choose from, and for the second prize we have a set of 9 kids, so by the product rule, the number of ways that 2 prizes can be given without giving both to the same kid is $10 \cdot 9 = 90$.

Problem 10. In how many ways can 2 prizes be given to a class of 10 kids, if it is allowed that both be given to the same kid?

Answer. The first prize can be given to one in a set of of 10 kids, and the second prize can also be given to one in a set of 10 kids, so by the product rule, the number of ways that 2 prizes can be given if it is allowed that both be given to the same kid is $10 \cdot 10 = 100$.

Problem 11. With 20 forks and 24 spoons available, how many ways can a man choose a fork and a spoon? Then how can another man take another fork and spoon?

Answer. The first man can choose a fork and spoon in $20 \cdot 24 = 480$ ways, and the second man can choose a fork and spoon in $19 \cdot 23 = 437$ ways.

Problem 12. From a list of 9 Corvidae birds, and 48 Columbidae birds, in how many ways can we choose an example of each?

Answer. $9 \cdot 48 = 432$ ways.

Problem 13. From 12 masculine words, 9 feminine, and 10 neuter, in how many ways can we choose an example of each?

Answer. $12 \cdot 9 \cdot 10 = 1,080$ ways.

Problem 14. A friend shows me 5 physics books, 7 math books, and 10 history books, and allows me to choose 2 books, on the condition that they must not be both of the same subject. How many selections can I make?

Answer. I can apply the product rule to each possible subject pair, so I can choose a physics and a math book in $5 \cdot 7 = 35$ ways, a physics and a history book in $5 \cdot 10 = 50$ ways, and a math and a history book in $7 \cdot 10 = 70$ ways. And by the sum rule, I can make a total of $35 + 50 + 70 = 155$ selections.

Problem 15. In how many ways can we select a consonant and a vowel, if there are 20 consonants and 6 vowels?

Answer. The total number of consonant-vowel pairs we can select is $20 \cdot 6 = 120$.

Problem 16. In how many ways can we make a 2 letter word consisting of one consonant and one vowel?

Answer. From the previous answer, we have 120 possible consonant-vowel pairs. Each consonant-vowel pair can be arranged in 2 ways to form a word. So the total number of 2 letter words is $120 \cdot 2 = 240$.

Problem 17. With nine pairs of gloves, in how many ways can I choose a left hand glove and a right hand glove, so that the resulting pair is not a matching pair?

Answer. If I choose the left hand glove first, I have a set of 9 to choose from, then for the right hand glove I have a set of 8 to choose from, because the matching right hand glove is not included. So by the product rule, I can choose $9 \cdot 8 = 72$ possible pairs. Or... I can choose from 9 left hand gloves, and 9 right hand gloves, which gives $9 \cdot 9 = 81$ choices, but then I must subtract the 9 possible matching pairs, which leaves $81 - 9 = 72$ choices, as before.

Problem 18. With 6 pairs of unique gloves, how many

ways can you take a left hand glove and a right hand glove that are unmatched?

Answer. A left hand glove can be chosen in 6 ways, and an unmatched right hand glove can be chosen in 5 ways, so there are a total of $6 \cdot 5 = 30$ ways.

Problem 19. From 7 unicycles, 5 bicycles, and 8 skateboards, how many ways can I select one of each?

Answer. $7 \cdot 5 \cdot 8 = 280$ ways.

Problem 20. Two people get into a bus that has 6 empty seats, in how many ways can they be seated?

Answer. The first person to board has a set of 6 seats to choose from, the second person has a set of 5 seats to choose from, so by the product rule, they can be seated in $6 \cdot 5 = 30$ different ways. The answer is also gotten by asking "How many permutations are there of 6 distinct items taken 2 at a time?" which is simply $\frac{6!}{(6-2)!} = 30$.

Problem 21. How many ways can we make a 2 letter word, out of an alphabet of 26 letters, if the 2 letters must be different?

Answer. For the first letter we have 26 letters to choose from, while for the second letter we have 25 letters to choose from, so the total number of possible words is $26 \cdot 25 = 650$ ways. The answer is also the number of permutations of 26 distinct things taken 2 at a time $= \frac{26!}{(26-2)!} = 650$.

Problem 22. There are 6 bibles, 3 prayer books, and 4 hymn books, in addition to 5 volumes of a bible and prayer book bound together, and 7 volumes of a prayer and hymn book bound together. In how many ways can I select a bible, a prayer book, and a hymn book?

Answer. As singles, there are $6 \cdot 3 \cdot 4 = 72$ ways to select, using a bible + prayer book bound volume there are $5 \cdot 4 = 20$ ways, and using a prayer + hymn book bound volume there are $7 \cdot 6 = 42$ ways, for a total of $72 + 20 + 42 = 134$ ways.

Problem 23. If there were also 3 volumes of a bible and hymn book bound together, how many ways would there now be?

Answer. In addition to the above 134 ways, I could also use a bible + hymn book bound volume for $3 \cdot 3 = 9$ more ways, for a total of $134 + 9 = 143$ ways.

Problem 24. A bowl holds 12 mangos and 10 grapefruits, Spike will choose a mango *or* a grapefruit, and then Spud will choose a mango *and* a grapefruit. Show that if Spike chooses a mango, then

Spud will have more ways to choose than if Spike chose a grapefruit.

Answer. If Spike chooses a mango, then Spud will have $11 \cdot 10 = 110$ ways to choose a mango and a grapefruit. On the other hand, if Spike chooses a grapefruit then Spud will have $12 \cdot 9 = 108$ ways to choose. Therefore Spud has more ways to choose, if Spike chooses a mango.

Problem 25. There are 12 ladies and 10 gentlemen, of whom 3 ladies and 2 gentlemen are brothers and sisters, the rest being unrelated. How many ways can there be an unrelated marriage?

Answer. Neglecting relations, there are $12 \cdot 10 = 120$ possible marriages, but of those, there are $3 \cdot 2 = 6$ related pairs, so that leaves $120 - 6 = 114$ unrelated marriages.

Problem 26. In how many ways can 8 different bells be rung? If one of the bells must always be rung last how many ways can they all be rung?

Answer. There are 8 ways to chose the first bell, 7 ways to chose the second, and so on. The number of ways to ring the bells is then $8! = 40,320$. This is the number of permutations of 8 distinct things. If one bell is always rung last then the others can be rung in $7! = 5,040$ ways.

Problem 27. In how many ways can the following set of letters be arranged in a row: $a, a, a, a, a, a, a, b, c$?

Answer. If we start by putting the 7 identical a's in a row with spaces between them:

$$a \ a \ a \ a \ a \ a \ a$$

then the b can be placed in any of the 6 spaces or on either end, making 8 possible places. Once the b is placed the letter c can be placed in any of the 7 remaining spaces or on either side of the b, making 9 possible places. This results in a total of $8 \cdot 9 = 72$ possible arrangements.

Another way to solve this problem is to imagine that the a's are somehow uniquely labeled. We then have 9 unique letters that can be arranged in 9! ways. If we now remove the labels then a given arrangement will be duplicated 7! times. The number of unique arrangements is then $9!/7! = 9 \cdot 8 = 72$.

Problem 28. The Local Galactic Congress consists of 8 humans, 13 Alpha Centaurians, 21 Lalandians, 11 Eridanians, 19 Aquarions, 10 Cygnians, and 15 Tau Cetians. In how many ways is it possible to form a committee of 7 members representing each of the 7 stellar life groups?

Answer. $8 \cdot 13 \cdot 21 \cdot 11 \cdot 19 \cdot 10 \cdot 15 = 68,468,400$.

Problem 29. Twenty people compete in a race with 3 prizes. In how many ways can the prizes be awarded?

Answer. The first prize can be awarded 20 ways, the second 19 ways, and the third 18 ways, so the total number of ways is $20 \cdot 19 \cdot 18 = 6,840$. Or, it is simply the number of permutations of 20 objects taken 3 at a time $= \frac{20!}{(20-3)!} = 6,840$.

Problem 30. How many ways can 4 letters be put into 4 envelopes, with only one letter per envelope?

Answer. For the first letter there are 4 envelopes to choose from, for the second letter there are 3 envelopes, for the third, 2 envelopes, and for the fourth, one. The total number of ways is then $4 \cdot 3 \cdot 2 \cdot 1 = 24$. It can also be stated as the number of permutations of 4 distinct items taken 4 at a time $= \frac{4!}{(4-4)!} = 24$.

Problem 31. How many ways can you sum a silver dollar, a quarter, a dime, a nickel, and a penny?

Answer. Each coin can either be summed or not, and there are 5 coins, so the ways you can sum them is $2^5 = 32$. But this includes the case of not summing any coin, which isn't really summing them, so the answer is 31.

Problem 32. There are 20 candidates for an office, and 7 voters. How many ways can the votes be made?

Answer. Each of the 7 voters can choose one of 20 candidates, so by the product rule, the number of ways is $20^7 = 1,280,000,000$.

Problem 33. I have 6 letters to be delivered to different parts of town, and two kids offer their services to deliver them. In how many different ways can I choose to send the letters?

Answer. I can choose to give the first letter to one of either kids, the second letter to one of either kids, and so on for all 6 letters. Therefore there are 2 choices for each letter, so the number of ways is $2^6 = 64$.

Problem 34. In how many ways can 6 different things be divided between 2 kids?

Answer. This question appears to be identical to the previous, so the answer should be 64. But the answer to the previous question included the case of one kid delivering all the letters, and the other kid having none to deliver, which isn't really a way to divide the letters between the 2 kids. Therefore the answer is $64 - 2 = 62$.

Problem 35. In how many ways can six different things be divided into two parcels?

Answer. This question seems identical to the previous, but that question made a distinction between the two kids, whereas with parcels, we don't. For example the first kid could have 4 things, and the second kid 2. This was distinguished from the first kid having 2 things, and the second kid 4. With parcels there is no such distinction, so the answer is not 62 like previously, but half of that, 31.

Problem 36. In how many ways can the following fruits be divided between 2 people: *fig, fig, fig, fig, date, date, date, kiwi, kiwi, lime?*

Answer. Considering only how the 4 figs can be divided, one person can get none, or just one, or 2, 3, or 4. So there are 5 ways to divide the 4 figs. Similarly there are 4 ways to divide the 3 dates, 3 ways to divide the 2 kiwis, and 2 ways to divide the single lime. So all the fruits can be divided in $5 \cdot 4 \cdot 3 \cdot 2 = 120$ ways. But this includes the case of one person getting all the fruits, and the other getting none, which isn't really a way to divide them, so excluding those 2 cases, the answer is 118.

Problem 37. There are 3 teetotums having 6, 8, and

10 sides. In how many ways can they fall? And of those ways, how many will have at least two 1's show up?

Answer. They can fall in $6 \cdot 8 \cdot 10 = 480$ ways. There are 4 categories of ways to get at least two 1's. The first category is to get three 1's, for which there is only one way. The second category is for the first 2 teetotums to have 1's and the last anything but a 1, for which there are 9 ways. The third category is for the first and third teetotum to have 1's, and the second anything but a 1, for which there are 7 ways. The fourth category is for the second and third teetotums to have 1's, and the first anything but a 1, for which there are 5 ways. So the total number of ways is $1+9+7+5 = 22$.

Problem 38. Having cloth of 5 different colors, in how many ways can I choose 3 colors for a tricolor flag?

Answer. $\binom{5}{3} = \frac{5!}{3! \cdot 2!} = 10$ ways.

Problem 39. In how many ways can I arrange vertically the 3 colors as 3 horizontal strips?

Answer. This is the number of permutations of 3 unique things taken 3 at a time $= \frac{3!}{(3-3)!} = 6$ ways. So

the total number of different flags I can make is
$10 \cdot 6 = 60$.

Problem 40. In the ordinary system of notation, how
many numbers are there which consist of five digits?

Answer. The leftmost digit can have any of the 10
numerals except for 0, while the remaining 4 digits can have any of the 10 numerals. So the total
number of numbers which consist of 5 digits is
$9 \cdot 10 \cdot 10 \cdot 10 \cdot 10 = 90,000$.

Problem 41. If a password consists of 4 letters, each
letter chosen from 26 possibilities, how many different possible passwords are there?

Answer. $26^4 = 456,976$.

Problem 42. I would like to publish a set of dictionaries to translate from one language to any other.
If I limit myself to 5 languages, how many dictionaries do I need to publish?

Answer. This is the number of permutations of 5 unique
things taken 2 at a time $= \frac{5!}{(5-2)!} = 20$ dictionaries.

Problem 43. If I extend my limit to 10 languages,
how many more dictionaries do I need to publish?

Answer. For 10 languages I need to publish $\frac{10!}{(10-2)!} = 90$ dictionaries, which is 70 more dictionaries than if I had only 5 languages.

Problem 44. A father lives near 3 boys schools and 4 girls schools. In how many ways can he send his 3 sons and 2 daughters to school?

Answer. $3^3 \cdot 4^2 = 432$ ways.

Problem 45. With 300 names to choose from, in how many ways can a child be named if we don't use more than 3 names at once?

Answer. Using only one name, there are 300 choices. Using 2 names there are $\frac{300!}{(300-2)!} = \frac{300!}{298!} = 89,700$ choices, and using 3 names there are $\frac{300!}{(300-3)!} = \frac{300!}{297!} = 26,730,600$ choices. So the total number of ways a child can be named is $26,730,600 + 89,700 + 300 = 26,820,600$.

Problem 46. In how many ways can we arrange 6 statues in 6 niches, one in each?

Answer. For the first statue there are 6 niches to choose from, for the second statue, 5 niches, for the third statue, 4 niches, for the fourth statue, 3 niches, for the fifth statue, 2 niches, and for the sixth statue 1 niche. So the total number

of ways we can arrange 6 statues in 6 niches is
$6! = 6 \cdot 5 \cdot 4 \cdot 3 \cdot 2 \cdot 1 = 720$.

Problem 47. In how many ways can 12 ladies and
12 gentlemen form themselves into couples for a
dance?

Answer. The first gentleman has 12 ladies to choose
from, the second gentleman has 11 ladies to choose
from, the third gentleman has 10 ladies to choose
from, and continuing until the twelfth gentleman
has one lady to choose. Therefore 12 ladies and
12 gentlemen can form themselves into $12! =
12 \cdot 11 \cdot 10 \ldots 3 \cdot 2 \cdot 1 = 479,001,600$ couples.

Problem 48. With 12 ladies and 12 gentlemen in a
ballroom, in how many ways can they take their
places for a contra dance (couples dance in 2 fac-
ing lines)?

Answer. From the previous question, there are 12!
different possible couples, and the 12 couples can
be arranged in a line in 12! different ways. So the
total number of ways they can take their places
for a contra dance is
$12! \cdot 12! = 229,442,532,802,560,000$.
Alternatively, the gentlemen can arrange themselves in a line in 12! different ways, and the ladies
can arrange themselves in a line in 12! different

ways. So the total number of ways they can take their places is $12! \cdot 12!$ as we just found.

Problem 49. In how many different ways can the letters a, b, a, d, e, f be arranged so that each arrangement begins with ab?

Answer. If each arrangement begins with ab, it's as if we simple removed those 2 letters from the arrangements, so that we only have the remaining 4 letters: a, d, e, f, and 4 items can be arranged in $4! = 24$ ways.

Problem 50. A shelf holds 5 biology books, 6 history, and 8 math books. In how many ways can the 19 books be arranged, keeping all the biology together, all the history together, and all the math together?

Answer. The 5 biology books can be arranged in $5! = 120$ ways, the 6 history books in $6! = 720$ ways, and the 8 math books in $8! = 40,320$ ways. The 3 different categories of books can be arranged in $3! = 6$ ways. So the total number of possible arrangements is the product of these ways: $120 \cdot 720 \cdot 40,320 \cdot 6 = 20,901,888,000$ ways.

Problem 51. In how many ways can the same books be arranged indiscriminately on the shelf?

Answer. 19! = 121, 645, 100, 408, 832, 000 ways.

Problem 52. A table is set for 8 people. How many ways can they be seated?

Answer. 8! = 40, 320 ways.

Problem 53. In how many arrangements can 4 people sit at a round table so that none will have the same neighbors?

Answer. If I am one of the 4 people, the other 3 people can be arranged in 3! = 6 ways, but half of those ways are the same arrangements with respect to rotation which doesn't change the neighbors. So the number of arrangements is $\frac{3!}{2} = 3$.

Problem 54. In how many ways can 7 people sit as described in the previous question? In how many of these ways will 2 assigned people be neighbors? In how many of these ways will an assigned person have the same 2 neighbors?

Answer. If I am one of the 7 people, the other 6 can be arranged in 6! = 720 ways, but half of those have the same neighbors, so the number of arrangements is $\frac{720}{2} = 360$. Out of these 360 arrangements I have 720 neighbors, and if I am one of 2 assigned people, the other assigned person must be my neighbor $\frac{720}{6} = 120$ times. For the case of an assigned person having the same 2 neighbors, there are $4! = 24$ arrangements of the other 4 people, with 2 arrangements of the same neighbors, but half of all these arrangements have the same neighbors, so the result is $\frac{24 \cdot 2}{2} = 24$ ways.

Problem 55. How many ways can 8 children form themselves in a ring to dance around a maypole?

Answer. In this case, relative position is all that matters. Take for example the 4 arrangements, where the 8 children are represented by letters A through H.

```
    A   B         C   D         D   E         A   H
H                C   B       E   C       F   B           G
G                D   A       F   B       G   C           F
    F   E         H   G         A   H         D   E
```

If we first place A, and all the others relative to it, we see that the first 3 arrangements above are identical, but the last is different (having its circularity reversed). Since relative position is all that matters, then the number of possible arrangements are to be counted after the first is placed, leaving $7! = 5,040$ ways.

Problem 56. In how many ways can eight beads be strung on an elastic band to form a bracelet?

Answer. This question appears to be identical to the last, but there is a difference. For the 4 arrangements shown in the previous question, the rightmost arrangement is different for children around

⁎ a maypole, but it's the same for a bracelet, since it only represents a rotation of the bracelet, and not a difference in its construction. By this reasoning, for a bracelet there are only half as many unique arrangements as for children around a maypole. This makes $\frac{7!}{2} = 2,520$ ways.

Problem 57. How many 3 letter words can be made from an alphabet of 26 letters, with each letter used only once?

Answer. $26 \cdot 25 \cdot 24 = 15,600$ words.

Problem 58. How many 4 letter words?

Answer. $26 \cdot 25 \cdot 24 \cdot 23 = 358,800$ words.

Problem 59. How many 8 letter words?

Answer. $26 \cdot 25 \cdot 24 \cdot 23 \cdot 22 \cdot 21 \cdot 20 \cdot 19 = 62,990,928,000$ words.

Problem 60. 5 women and 3 men get together to play croquet. In how many ways can they divide themselves into 2 groups of 4 so that the men are not all on the same side?

Answer. There is only one way to split the men into 2 groups so that they're not on the same side, 2 on one side, and one on the other. With 3 men, this can be done in $\binom{3}{2} = \frac{3!}{2!} = 3$ ways. The side with 2 men on it has room for 2 women, and the side with one man on it has room for 3 women, so the women can be split in $\binom{5}{2} = \frac{5!}{2!\cdot 3!} = 10$ ways. Thus the total number of ways is $3 \cdot 10 = 30$.

Problem 61. I have 6 packages to be delivered with a choice of 3 carriers. In how many ways can I send the packages?

Answer. For each package, I have a choice of 3 carriers, so the total number of ways is $3^6 = 729$.

Problem 62. Spike has 7 different books, while Spud has 9 different books. In how many ways can one of Spike's books be exchanged for one of Spud's?

Answer. Spike can choose one of his 7 books to exchange, while Spud can choose one of his 9, so there are $7 \cdot 9 = 63$ ways.

Problem 63. For the previous question, how many ways can 2 of Spike's books be exchanged for 2 of Spud's?

Answer. Spike can choose 2 out of 7 of his books, which makes $\binom{7}{2} = \frac{7!}{2!\cdot 5!} = 21$ ways. Spud can choose 2 out of 9, which makes $\binom{9}{2} = \frac{9!}{2!\cdot 7!} = 36$

ways. The total number of ways is then $21 \cdot 36 = 756$.

Problem 64. Five guys, Spike, Spud, Skeeter, Sparky, and Spivy will be speaking at a meeting. How many ways can they take their turn without Spud speaking before Spike?

Answer. Without restrictions, the total number of ways the men can speak is $\frac{5!}{(5-5)!} = 120$. In half of these ways Spike will speak before Spud, and vice versa for the other half. So there are $\frac{120}{2} = 60$ ways without Spud speaking before Spike.

Problem 65. For the previous question, in how many ways can Spike speak immediately before Spud?

Answer. If we consider Spike and Spud to be one speaker, then there are $\frac{4!}{(4-4)!} = 24$ ways they can be arranged with the other 3 speakers. This is the number of ways Spike can speak immediately before Spud.

Problem 66. 4 flags are to be raised on one mast, with 20 different flags to choose from. How many different ways can the flags be arranged?

Answer. $20 \cdot 19 \cdot 18 \cdot 17 = 116,280$ ways. The answer would be the same if there were 4 different masts for the 4 flags, since there are still only 4 positions.

Problem 67. A boat with 8 oars has to be manned by a club with 50 members. How many ways can we arrange the crew?

Answer. $50 \cdot 49 \cdot 48 \cdot 47 \cdot 46 \cdot 45 \cdot 44 \cdot 43 = 21,646,947,168,000$ ways.

Problem 68. In how many ways can 2 people divide 30 books between them, with one having twice as many as the other?

Answer. This is the number of combinations of 30 taken 20 at a time, or $\binom{30}{20} = \frac{30!}{20! \cdot 10!} = 30,045,015$ ways.

Problem 69. There are 8 men to row an 8 oared boat, but 2 of them can only row on stroke side, and one of them only on bow side, with the rest able to row on either side. How many possible arrangements of the men are there?

Answer. Of the 5 remaining men who can row on either side, 2 of them are needed for stroke side, and 3 on bow side. These 5 men can therefore be arranged in $\binom{5}{2} = \frac{5!}{2! \cdot 3!} = 10$ ways. The 4 men on stroke side can be arranged in $4! = 24$ ways, and the 4 men on bow side can similarly be arranged in 24 ways. So the total number of possible arrangements is $10 \cdot 24 \cdot 24 = 5,760$ ways.

Problem 70. In a company of soldiers there are 3 officers, 4 sergeants, and 60 privates. In how many ways can we form a detachment consisting of an officer, 2 sergeants, and 20 privates? In how many of those ways will the captain and senior sergeant appear?

Answer. Such a detachment can be formed in $\binom{3}{1} \cdot \binom{4}{2} \cdot \binom{60}{20} = 3 \cdot 6 \cdot 4,191,844,505,805,495 = 75,453,201,104,498,910$ ways. Of these ways, the captain appears once every 3 times, and the senior sergeant appears half the time. So the number of ways the captain and senior sergeant appear is the total ways above divided by 6 or $\frac{75,453,201,104,498,910}{6} = 12,575,533,517,416,485$.

Problem 71. From 12 ladies and 15 gents, how many ways can 4 ladies and 4 gents be chosen for a dance?

Answer. $\binom{12}{4} \cdot \binom{15}{4} = 495 \cdot 1,365 = 675,675$ ways.

Problem 72. A guy belongs to a fruit-of-the-month club which has 30 fruits available to choose from.

If he selects 5 fruits a month, forming a different group of 5 each month, for how many months can he do this?

Answer. $\binom{30}{5} = 142,506$ months, or 11,875.5 years.

Problem 73. In how many ways can 3 kids divide 12 oranges, the oranges all being of different sizes?

Answer.

$$\binom{12}{4,4,4}$$
$$= \frac{12!}{4! \cdot 4! \cdot 4!}$$
$$= 34,650$$

ways.

Problem 74. In how many ways can they divide them so that the eldest gets 5, the middle 4, and the youngest 3?

Answer.

$$\binom{12}{5,4,3}$$
$$= \frac{12!}{5! \cdot 4! \cdot 3!}$$
$$= 27,720$$

ways.

Problem 75. If there are 15 identical apples, 20 identical pears, and 25 identical oranges, in how many ways can 60 kids take one each?

Answer.

$$\binom{60}{15, 20, 25}$$
$$= \frac{60!}{15! \cdot 20! \cdot 25!}$$
$$= 168, 618, 391, 667, 123, 831, 595, 882, 720$$
$$\approx 1.7 \text{x} 10^{26}$$

ways.

Problem 76. Using 6 dice, in how many ways can you throw 2 sixes, 3 fives, and a one?

Answer.

$$\binom{6}{2, 3, 1} = \frac{6!}{2! \cdot 3! \cdot 1!} = 60$$

ways.

Problem 77. Show that the letters of *anticipation* can be arranged in 3 times as many ways as the letters of *commencement*.

Answer. *anticipation* has 12 letters, with 2 *a*'s, 2 *n*'s, 2 *t*'s, 3 *i*'s, and single occurrences of the other 3 letters. The letters of this word can be arranged in $\binom{12}{2,2,2,3} = 9,979,200$ ways. *commencement* has 12 letters, with 2 c's, 3 m's, 3 e's, 2 n's, and single occurrences of the other 2 letters. The letters of this word can be arranged in $\binom{12}{2,3,3,2} = 3,326,400$ ways, which is $\frac{1}{3}$ the number of ways for the previous word.

Problem 78. How many 5 letter words can be made from 26 letters with repetitions allowed, but with no 2 identical letters as neighbors?

Answer. The first letter can be chosen in 26 ways, the second in 25 ways (omitting the letter just chosen), the third in 25 ways (omitting the letter just chosen), etc. So the number of such 5 letter words is $26 \cdot 25^4 = 10,156,250$.

Problem 79. 8 people will row a boat, but 2 can only row on the left side, and three only on the right side, with the rest able to row on either side. In how many ways can the people be arranged?

Answer. Of the 3 remaining people who can row on either side, 2 of them are needed for the left side, and one on the right side. These 3 people can therefore be arranged in $\binom{3}{2} = 3$ ways. The 4 people on the left side can be arranged in $4! = 24$ ways, and the 4 people on the right side also in 24 ways. So the total number of possible arrangements is $3 \cdot 24 \cdot 24 = 1,728$.

Problem 80. In how many ways can a deck of 52 cards be divided among 4 players so that each gets 13 cards?

Answer.

$$\binom{52}{13, 13, 13, 13}$$
$$= \frac{52!}{13! \cdot 13! \cdot 13! \cdot 13!}$$
$$= 53,644,737,765,488,792,839,237,440,000$$
$$\approx 5.4\text{x}10^{28}$$

ways.

Problem 81. In how many ways can a deck of 52 cards be divided into 4 groups of 13 cards?

Answer. This question is different from the previous only by the fact that the 4 players in the previous question were distinguishable, whereas the 4 groups in this problem are not. The consequence is that the answer to the previous question must be divided by 4!. So we have

$$\binom{52}{13, 13, 13, 13} \cdot \frac{1}{4!}$$
$$= \frac{52!}{13! \cdot 13! \cdot 13! \cdot 13! \cdot 4!}$$
$$= 2,235,197,406,895,366,368,301,560,000$$
$$\approx 2.2 \text{x} 10^{27}$$

ways.

Problem 82. In how many ways can 12 books be divided into 3 groups, with one group having 3 members, another 4 members, and the last 5 members?

Answer. This question is different from the previous by the fact that the 3 groups in this question have different numbers of members, with the consequence that the groups are distinguishable, therefore the number of ways is $\binom{12}{3,4,5} = 27,720$.

Problem 83. I have 3 copies of Tarzan, 2 copies of Robinson Crusoe, and 1 copy of Don Quixote. In how many ways can I give them to a class of 12 students, (1) so that no student gets more than 1 book and (2) so that no student receives more than 1 copy of any book?

Answer. (1) If no student gets more than 1 book, then there are $\binom{12}{3} \cdot \binom{9}{2} \cdot \binom{7}{1} = 220 \cdot 36 \cdot 7 = 55,440$ ways to give them. (2) If no student gets more than 1 copy of any book, then there are $\binom{12}{3} \cdot \binom{12}{2} \cdot \binom{12}{1} = 220 \cdot 66 \cdot 12 = 174,240$ ways.

Problem 84. In how many ways can a set of 12 black checkers and 12 white checkers be placed on the 32 black squares of a checker board?

Answer. We divide the 32 black squares of the checker board into 3 different subsets, the first subset of size 12 contains black checkers, the second subset of size 12 contains white checkers, and the remaining 8 are empty. The number of combinations of these 3 subsets is therefore $\binom{32}{12,12,8} = 28,443,124,054,800$.

Problem 85. Out of 100 things, how many ways can 3 things be selected?

Answer. $\binom{100}{3} = \frac{100!}{3! \cdot 97!} = 161,700$ ways.

Problem 86. From a basket of twenty pears, at 3 pears for a dollar, how many ways can you select 6 dollars worth?

Answer. With 6 dollars and at 3 pears for a dollar we can select 18 pears. The 18 pears can be selected in $\binom{20}{18} = \frac{20!}{18! \cdot 2!} = \frac{20 \cdot 19}{2} = 190$ ways.

Problem 87. In how many ways can we make the same choice, if we always take the largest pear?

Answer. Always choosing the largest pear, there are 17 additional pears to choose from the 19 remaining, so the number of ways is $\binom{19}{17} = \frac{19!}{17! \cdot 2!} = \frac{19 \cdot 18}{2} = 171$.

Problem 88. In how many ways can we make the same choice without taking the smallest pear?

Answer. Taking the smallest pear out of the running, there are 18 to choose from the 19 remaining, so the number of ways is $\binom{19}{18} = 19$.

Problem 89. In how many ways can we make the same choice by both including the largest pear, and not including the smallest pear?

Answer. After taking the largest pear, and removing the smallest pear from the running, there are 17 pears to choose from the 18 remaining. This results in $\binom{18}{17} = \frac{18!}{17!} = 18$ ways.

Problem 90. Out of 42 liberals and 50 conservatives, how many choices are there in selecting a committee of 4 liberals and 4 conservatives?

Answer. $\binom{42}{4} \cdot \binom{50}{4} = \frac{42 \cdot 41 \cdot 40 \cdot 39}{4!} \cdot \frac{50 \cdot 49 \cdot 48 \cdot 47}{4!} = 111,930 \cdot 230,300 = 25,777,479,000$ ways.

Problem 91. In how many ways can we arrange the letters of the word *indivisibility?*

Answer. If all 14 letters of this word were unique, the answer would be 14!. But there are 6 identical letters (the *i*'s). The result is that 14! is larger than the answer by a factor of 6!, so the number of ways is $\frac{14!}{6!} = 121,080,960$.

Problem 92. In how many ways can we arrange the letters of the word *parallelepiped?*

Answer. By similar reasoning to the previous problem, we have 14 letters total, but the duplicates are 3 *p*'s, 2 *a*'s, 3 *l*'s, and 3 *e*'s, so the number of ways is $\frac{14!}{3!\cdot2!\cdot3!\cdot3!} = 201,801,600$.

Problem 93. In how many ways can we arrange the letters of the word *llangollen?*

Answer. $\frac{10!}{4!\cdot2!} = 75,600$ ways.

Problem 94. In how many ways can the letters of the word *possessions* be arranged?

Answer. This word has 11 letters, with 2 *o*'s, 5 *s*'s, and single occurrences of the remaining letters. So the letters of this word can be arranged in $\binom{11}{2,5} = 166,320$ ways.

Problem 95. In how many ways can the letters of the word *cockatoo* be arranged?

Answer. This word has 8 letters, with 2 *c*'s, 3 *o*'s, and single occurrences of the remaining letters. So the letters of this word can be arranged in $\binom{8}{2,3} = 3,360$ ways.

Problem 96. Show that the letters of *cocoon* can be arranged in twice as many ways as the letters of *cocoa*.

Answer. *cocoon* has 6 letters, with 2 *c*'s, 3 *o*'s, and one *n*, which allows $\binom{6}{2,3} = 60$ arrangements. *cocoa* has 5 letters, with 2 *c*'s, 2 *o*'s, and one *a*, which allows $\binom{5}{2,2} = 30$ arrangements, and that's half the arrangements of the word *cocoon*.

Problem 97. In how many ways can you arrange the letters of *pallmall* without letting all the *l*'s come together?

Answer. *pallmall* has 8 letters with 2 *a*'s, 4 *l*'s, and single occurrences of the remaining 2 letters. Without restrictions, there are $\binom{8}{2,4} = 840$ arrangements. But there are $\binom{5}{2} = 60$ arrangements with all *l*'s together, so the number of arrangements without all the *l*'s together is $840 - 60 = 780$.

Problem 98. In how many ways can you arrange the letters
oiseau so that the vowels are in alphabetical order?

Answer. This word contains 5 vowels and 1 consonant. If we put the vowels in alphabetical order *aeiou* then the consonant *s* is the only other letter to be arranged. It can be placed between these vowels or on the ends, or 6 ways.

Problem 99. In how many ways can you arrange the letters of *cocoa* so that *a* is in the middle?

Answer. With *a* in the middle, there are 2 letters on either side, that are selected from a pair of *c*'s and a pair of *o*'s. This selection can be done in $\binom{4}{2,2} = 6$ ways.

Problem 100. In how many ways can you arrange the letters of *quartus* so that *q* is followed by *u*?

Answer. Keeping *qu* together, we can consider it one letter, so there are 6 letters instead of 7. The number of arrangements then are $\frac{6!}{(6-6)!} = \frac{6!}{0!} = 720$.

Problem 101. In how many ways can you arrange the letters of *ubiquitous* so that *q* is followed by *u*?

Answer. Keeping *qu* together, we can consider it one letter, so there are 9 letters instead of 10. Of these 9 letters, there are 2 *u*'s, 2 *i*'s, and the rest occur only once. The number of arrangements then are $\binom{9}{2,2} = 90,720$.

Problem 102. In how many ways can you arrange the letters of *quisquis* so that each *q* is followed by *u*?

Answer. Keeping both *qu* pairs together, we can consider each pair to be a letter, so instead of 8 letters, there are 6. Of these 6 letters, there are 2 *qu*'s, 2 *i*'s, and 2 *s*'s. The number of arrangements then are $\binom{6}{2,2,2} = 90$.

Problem 103. In how many ways can you arrange the letters of *indivisibility* without letting 2 *i*'s be together?

Answer. This word contains 14 letters, 6 of which are *i*'s, the remaining 8 occurring only once. The remaining 8 letters can be arranged in $8! = 40,320$ ways. For each of these arrangements the 6 *i*'s can be placed in between the 8 letters or on either end, for 9 possible positions. The number of ways this can be done is $\binom{9}{6} = 84$. So the total number of ways is $40,320 \cdot 84 = 3,386,880$.

Problem 104. In how many ways can you arrange the letters of *facetious* without letting 2 vowels be together?

Answer. *facetious* contains 9 letters, with 5 vowels and 4 consonants. The 4 consonants can be arranged in $4! = 24$ ways. The 5 vowels, which are all unique, can be placed in between the 4 consonants or on the ends, making 5 possible positions. This can be done in $\frac{5!}{(5-5)!} = \frac{5!}{0!} = 120$ ways. The total number of ways is then $24 \cdot 120 = 2,880$.

Problem 105. In how many ways can you rearrange the letters of *facetious* without changing the order of the vowels?

Answer. The 9 letters of *facetious*, which are all unique, can be arranged in $\frac{9!}{(9-9)!} = 9!$ ways. The 5 vowels can be arranged in $\frac{5!}{(5-5)!} = 5!$ ways. The number of arrangements of the 9 letters, with the vowels in their original order, is then $\frac{9!}{5!} = 3,024$. One of these arrangements is just the original word, so the number of rearrangements is 3,023.

Problem 106. In how many ways can you rearrange the letters of *abstemiously* without changing the order of the vowels?

Answer. The 12 letters of *abstemiously*, which are all unique except for 2 *s*'s, can be arranged in

$\binom{12}{2,1,1,\ldots,1}$ = 239,500,800 ways. The 5 vowels of this word can be arranged in $\frac{5!}{(5-5)!}$ = 5! ways. The number of arrangements of the 12 letters, with the vowels in their original order, is then $\frac{239,500,800}{5!}$ = 1,995,840. One of these arrangements is just the original word, so the number of rearrangements is 1,995,839.

Problem 107. In how many ways can you rearrange the letters of *parallelism* without changing the order of the vowels?

Answer. The 11 letters of *parallelism*, which are all unique except for 2 *a*'s and 3 *l*'s, can be arranged in $\binom{11}{2,3}$ = 3,326,400 ways. The 4 vowels of this word can be arranged in $\binom{4}{2,1,1}$ = 12 ways. The number of arrangements of the 11 letters, with the vowels in their original order, is then $\frac{3,326,400}{12}$ = 277,200. One of these arrangements is just the original word, so the number of rearrangements is 277,199.

Problem 108. In how many ways can you rearrange the letters of *almost*, maintaining the current separation of the vowels?

Answer. *almost* has 6 letters, with 4 consonants and 2 vowels. The 4 consonants can be arranged in $\frac{4!}{(4-4)!}$ = 4! = 24 ways. The 2 vowels, being in the first and fourth positions, can be arranged in 2 ways, but they can also occupy the second and fifth positions, as well as the third and sixth positions, so the total number of arrangements are $24 \cdot 2 \cdot 3 = 144$ ways. One of these ways is the original, so there are 143 ways to rearrange.

Problem 109. In how many ways can you rearrange the letters of *logarithms*, so that consonants occupy the second, fourth, and sixth positions (where the vowels currently are)?

Answer. *logarithms* has 10 letters, with 7 consonants and 3 vowels. The 3 positions where the consonants must be, can be filled in $\frac{7!}{(7-3)!} = \frac{7!}{4!} = 210$ ways. The remaining 7 positions can be filled with any of the remaining 7 letters, for $\frac{7!}{(7-7)!} = \frac{7!}{0!} = 5,040$ ways. So the total arrangements are $210 \cdot 5,040 = 1,058,400$ ways.

Problem 110. In how many ways can 2 consonants and a vowel be chosen from the word *logarithms*, and how many of those ways has the letter *s* in it?

Answer. 2 consonants can be chosen from the 7 in $\binom{7}{2} = \frac{7!}{2! \cdot 5!} = 21$ ways. A vowel can be chosen from the 3 in $\binom{3}{1} = \frac{3!}{2!} = 3$ ways. So the 2 consonants and a vowel can be chosen in $21 \cdot 3 = 63$

ways. If s is always chosen as one of the consonants, then there are 6 choices for the second consonant, and 3 choices for the vowel, for a total of $6 \cdot 3 = 18$ ways.

Problem 111. In how many ways can you arrange the letters of *syzygy* without all 3 y's being bunched together?

Answer. Without the restriction, there are $\binom{6}{3,1,1,1} = \frac{6!}{3!} = 120$ ways. The number of ways all 3 y's can be bunched together is $\frac{4!}{(4-4)!} = \frac{4!}{0!} = 24$. So the number of arrangements without all 3 y's bunched together is $120 - 24 = 96$ ways.

Problem 112. For the previous question, what is the number of ways without any 2 y's being bunched together?

Answer. The 3 letters s, z, and g can be ordered in $\frac{3!}{(3-3)!} = 6$ ways. The 3 y's can be placed in any of 4 positions around them, which can be done in $\binom{4}{3} = 4$ ways. So the total number of arrangements is $6 \cdot 4 = 24$.

Problem 113. In how many ways can you arrange the letters of the words *choice and chance* without any 2 c's being bunched together?

Answer. There are 15 letters, with 4 c's, 2 h's, 2 e's, 2 a's, 2 n's, and single occurrences for the remaining letters. Without the c's, there are 11 letters that can be arranged in $\binom{11}{2,2,2,2} = \frac{11!}{2^4} = 2,494,800$ ways. Between these 11 letters and on the ends, there are 12 positions we can place the 4 c's, with $\binom{12}{4} = 495$ ways to make these placements. The total number of ways to make the arrangements is then $2,494,800 \cdot 495 = 1,234,926,000$.

Problem 114. A company of volunteers consists of a captain, a lieutenant, an ensign, and 80 rank and file. In how many ways can 10 soldiers be selected so that the captain is always included?

Answer. Always choosing the captain, then there are 9 remaining to choose from 82 soldiers. So the number of ways is $\binom{82}{9} = \frac{82!}{73! \cdot 9!} = 293,052,087,900$.

Problem 115. How many ways can 10 soldiers be selected so that at least one officer is included?

Answer. There are $\binom{83}{10}$ ways to choose 10 soldiers from the 3 officers and the 80 rank and file. There are $\binom{80}{10}$ ways to choose 10 soldiers from the 80 rank and file. The difference $\binom{83}{10} - \binom{80}{10} = 785,840,219,450$ is the answer. We may also get the same result as follows. There are $3 \cdot \binom{80}{9}$ ways to choose one officer and the remaining 9 soldiers from rank and file. There are $3 \cdot \binom{80}{8}$ ways to choose 2 officers and the remaining 8 soldiers from rank and file. And there are $1 \cdot \binom{80}{7}$ ways to choose 3 officers and the remaining 7 soldiers from rank and file. The sum of these ways gives the same answer as above: $3 \cdot \binom{80}{9} + 3 \cdot \binom{80}{8} + 1 \cdot \binom{80}{7} = 785,840,219,450$.

Problem 116. How many ways can 10 soldiers be selected so that exactly one officer is included?

Answer. There are 3 ways to choose one officer, and $\binom{80}{9}$ ways to choose the 9 remaining soldiers from rank and file, therefore the number of ways is $3 \cdot \binom{80}{9} = 695,700,891,600$.

Problem 117. There are 15 candidates for admission into a society with 2 vacancies. There are 7 electors and each can either vote for one or 2 candidates. How many ways can the votes be made?

Answer. Each voter can choose a single candidate 15 ways, and 2 candidates $\binom{15}{2}$ ways, for a total of 120 ways. There are 7 voters, so the total number of ways the votes can be made is $120^7 = 358,318,080,000,000$.

Problem 118. Out of 20 men and 6 women, in how many ways can we choose 3 men and 3 women?

Answer. $\binom{20}{3} \cdot \binom{6}{3} = 1,140 \cdot 20 = 22,800$ ways.

Problem 119. Out of 20 men and 6 women, in how many ways can we fill 6 offices, with 3 requiring men and 3 requiring women?

Answer. This question is different from the previous in that the order of selection is important, so the number of ways is $\frac{20!}{(20-3)!} \cdot \frac{6!}{(6-3)!} = 20 \cdot 19 \cdot 18 \cdot 6 \cdot 5 \cdot 4 = 820,800$.

Problem 120. Using 20 consonants and 6 vowels, how many ways can we make a word with 3 different consonants and 3 different vowels?

Answer. We can choose 3 different consonants from 20 in $\binom{20}{3}$ ways, we can choose 3 different vowels from 6 in $\binom{6}{3}$ ways, and the number of permutations of the chosen 6 letters is 6!, so the total number of ways to make a word with 3 consonants and 3 vowels is the product of these ways: $\binom{20}{3} \cdot \binom{6}{3} \cdot 6! = \frac{20 \cdot 19 \cdot 18}{6} \cdot \frac{6 \cdot 5 \cdot 4}{6} \cdot 6 \cdot 5 \cdot 4 \cdot 3 \cdot 2 = 1,140 \cdot 20 \cdot 720 = 16,416,000$.

Problem 121. From the 26 letters of the alphabet, how many ways can we make a word with 4 different letters, where one of the letters must be a?

Answer. If one of the letters must be a, then there are 3 more letters to choose from the remaining 25 letters. The 3 letters can be chosen from the 25 in $\binom{25}{3} = \frac{25 \cdot 24 \cdot 23}{6} = 2,300$ ways. The 4 letters (including the a) can be arranged in $4! = 24$ ways, so the total number of ways we can make a word with 4 different letters, that includes an a, is $2,300 \cdot 24 = 55,200$. Alternatively, we can solve this problem by finding the number of 4 letter words, where the letters are different but there is no restriction on the a, as $26 \cdot 25 \cdot 24 \cdot 23 = 358,800$ words. The number of letters in these 4 letter words is $4 \cdot 358,800 = 1,435,200$. Now we note that since there was no bias in picking the letters, the number of a's in these 1,435,200 letters is $1,435,200/26 = 55,200$. This is also the number of words with a's in them, which matches the answer gotten previously.

Problem 122. From the 26 letters of the alphabet, how many ways can we make a 4 letter word, with 2 of them being a and b, and all of them being different?

Answer. Having already chosen a and b, we have 24 letters to choose 2 letters from, which can be done in $\binom{24}{2}$ ways. The number of ways to order 4 letters out of 4 is $4!$. So the total number of ways is $\binom{24}{2} \cdot 4! = 276 \cdot 24 = 6,624$.

Problem 123. From 20 consonants and 6 vowels, how many ways can we make a 5 letter word consisting of 2 different consonants and 3 different vowels, one of which must be a?

Answer. The 2 consonants can be chosen in $\binom{20}{2}$ ways, the 2 vowels (besides the a) can be chosen in $\binom{5}{2}$ ways, and the 5 letters can be ordered in 5! ways, so the total number of ways to make the 5 letter word are $\binom{20}{2} \cdot \binom{5}{2} \cdot 5! = 190 \cdot 10 \cdot 120 = 228,000$.

Problem 124. There are 10 jobs that need to be filled, 4 of which must be filled by men, 3 by women, and the remaining 3 by either. If there are 20 men and 6 women available, in how many ways can we fill the jobs?

Answer. The 20 men can fill the 4 jobs in $\frac{20!}{(20-4)!}$ ways, the 6 women can fill the 3 jobs in $\frac{6!}{(6-3)!}$ ways. There are 16 men and 3 women still available to fill the 3 remaining jobs, which can be done in $\frac{19!}{(19-3)!}$ ways. So the total number of ways to fill the 10 jobs is $\frac{20!}{(20-4)!} \cdot \frac{6!}{(6-3)!} \cdot \frac{19!}{(19-3)!} = 116,280 \cdot 120 \cdot 5,814 = 81,126,230,400$.

Problem 125. How many ways can you bundle 10 books into 5 parcels of 2 books each?

Answer. $\binom{10}{2,2,2,2,2} = \frac{3,628,800}{2^5} = 113,400$ ways. In this case the order of the 5 parcels doesn't matter, so we need to divide by 5!, which gives $\frac{113,400}{120} = 945$ ways.

Problem 126. How many ways can you bundle 9 books into 4 parcels of 2 books each, with one left over?

Answer. $\binom{9}{2,2,2,2,1} = \frac{362,880}{2^4} = 22,680$ ways. The order of the 4 parcels with 2 books each doesn't matter, so we need to divide by 4!, which gives $\frac{22,680}{24} = 945$ ways.

Problem 127. How many ways can you bundle 9 books into 3 parcels of 3 books each?

Answer. $\binom{9}{3,3,3} = \frac{362,880}{6^3} = 1,680$ ways. The order of the 3 parcels doesn't matter, so we need to divide by 3!, which gives $\frac{1,680}{6} = 280$ ways.

Problem 128. How many ways can 10 husband and wife couples be formed into 5 groups with 2 men and 2 women in each group?

Answer. This problem can be broken down into the number of ways of taking 10 distinct men and forming 5 distinct groups of 2 men each, and similarly for the women. For the men, this can be done in $\binom{10}{2,2,2,2,2} = \frac{10!}{2^5} = 113,400$ ways. The order of the 5 groups doesn't matter, so we must divide by 5!, giving $\frac{113,400}{120} = 945$ ways. The same result is gotten for the women in the same way. The number of ways to pair up the 5 groups of men with the 5 groups of women is 5!, so the total number of ways is $945 \cdot 945 \cdot 120 = 107,163,000$.

Problem 129. In how many of these ways will a given man find himself in the same group as his wife?

Answer. A given man has 2 women in his group. Since the choice of the 2 women out of 10 is unbiased, he will on average find himself in the same group as his wife $\frac{1}{5}$ of the time. So the number of ways is $\frac{107,163,000}{5} = 21,432,600$.

Problem 130. In how many of the ways will 2 given men find themselves in the same group as their wives?

Answer. If 2 given men and their wives make up one group, then the only variations are the number of ways that 8 men and 8 women can form 4 groups of 2 men and 2 women. This can be done in $\frac{\binom{8}{2,2,2,2}^2}{(4!)^2} \cdot 4! = 264,600$ ways (see the question before last for detailed explanation).

Problem 131. In how many ways can you select 6 bandanas, with repetitions allowed, at a shop that has 7 types available?

Answer. This question is equivalent to "How many multisets of size 6 can you create by sampling with replacement from a set of size 7?". The answer is $\binom{6+7-1}{7-1} = \binom{12}{6} = \frac{12!}{6! \cdot 6!} = 924$ ways.

Problem 132. From a 10 member choir, in how many ways can you select a different group of 6 every day for 3 days?

Answer. The number of subsets of size 6 from a set of size 10 is $\binom{10}{6} = \frac{10!}{6! \cdot 4!} = 210$. Selecting 3 items from 210 can be done in $\frac{210!}{(210-3)!} = \frac{210!}{207!} = 9,129,120$ ways.

Problem 133. A man has 6 friends, and invites 3 of them to dinner every day for 20 days. How many ways can he do this while selecting a different group of 3 each time?

Answer. The number of subsets of size 3 from a set of size 6 is $\binom{6}{3} = \frac{6!}{3! \cdot 3!} = 20$. Selecting 20 items from 20 can be done in $\frac{20!}{(20-20)!} = \frac{20!}{0!} = 2,432,902,008,176,640,000 \approx 2.43 \times 10^{18}$ ways.

Problem 134. From a large number of nickels, dimes, and quarters, in how many ways can 4 coins be selected?

Answer. This question is equivalent to asking "In how many ways can I put 4 items into 3 urns?". We can represent this with stars (items) and bars (urn boundaries). For example, one way to put 4 items into 3 urns is:

$$* \mid * \; * \mid *$$

This represents one item in the first urn, 2 items in the second urn, and one item in the third. The answer is the total number of ways of arranging these 6 items, with 4 of one kind (stars) and 2 of another (bars) in a line, or $\binom{6}{2} = \frac{6!}{2! \cdot 4!} = \frac{6 \cdot 5}{2} = 15$ ways. All the ways are shown below:

n n n n, n n d d, n d d d, n q q q, d d q q,
n n n d, n n d q, n d d q, d d d d, d q q q,
n n n q, n n q q, n d q q, d d d q, q q q q.

Problem 135. How many ways can we fill 3 glasses with 5 wines, without mixing?

Answer. Similar to the previous problem, this question can be posed as "In how many ways can I put 3 items into 5 urns?". Thinking in terms of "stars and bars" (see previous problem) the answer is the total number of ways of arranging 7 items, with 3 of one kind (stars) and 4 of another (bars) in a line, or $\binom{7}{3} = \frac{7!}{3! \cdot 4!} = \frac{7 \cdot 6 \cdot 5}{6} = 35$ ways. If we represent the 5 types of wine with vowels

(a, e, i, o, u), then the 35 ways are shown below:

$$aaa, \quad aee, \quad aio, \quad eee, \quad eio, \quad iii, \quad iuu,$$
$$aae, \quad aei, \quad aiu, \quad eei, \quad eiu, \quad iio, \quad ooo,$$
$$aai, \quad aeo, \quad aoo, \quad eeo, \quad eoo, \quad iiu, \quad oou,$$
$$aao, \quad aeu, \quad aou, \quad eeu, \quad eou, \quad ioo, \quad ouu,$$
$$aau, \quad aii, \quad auu, \quad eii, \quad euu, \quad iou, \quad uuu.$$

This answer assumes the glasses are indistinguishable, so that *aei* is the same as *aie*, *eai*, *eia*, *iae*, and *iea*. If the glasses are unique, then each can be filled in 5 ways, so the number of ways would be $5^3 = 125$.

Problem 136. In how many ways can a dozen marbles be selected in a shop where they sell 5 kinds of marbles?

Answer. Similar to the previous 2 problems, this question is equivalent to asking "In how many ways can I put 12 marbles into 5 urns?". So there are 12 stars and 4 bars, meaning the total number of ways to arrange 16 items with 12 of one kind and 4 of another is $\binom{16}{12} = \frac{16 \cdot 15 \cdot 14 \cdot 13}{4!} = 1,820$.

Problem 137. How many dominoes are there in a set numbered from double blank to double 9?

Answer. Each domino contains a selection of 2 numbers from 10 possibilities: 0, 1, 2, 3, 4, 5, 6, 7, 8,

9. Similar to the previous 3 problems, this question is equivalent to asking "In how many ways can I put 2 items into 10 urns?". So there are 2 stars and 9 bars, meaning the total number of ways to arrange 11 items with 2 of one kind and 9 of another is $\binom{11}{2} = \frac{11\cdot10}{2} = 55$.

Problem 138. In how many ways can an arrangement of 4 letters be made from letters of the words *choice and chance?*

Answer. There are 15 letters to choose from, with 8 unique ones. Duplicates are indicated in the following table. The ways the 4 letters can be

c	h	a	n	e	o	i	d
4	2	2	2	2	1	1	1

chosen falls into 5 categories:

1. Four all the same.
2. Three the same, one different.
3. Two the same, the other two the same.
4. Two the same, the other two different.
5. All four different.

Category 1 allows only one way, 4 c's. Category 2 requires 3 c's and one of the other 7 letters, with the other letter occupying one of 4 positions, so there are $7 \cdot 4 = 28$ ways. For category 3 we can choose the 2 pairs in $\binom{5}{2} = 10$ ways, and the 2 pairs can be arranged in $\binom{4}{2} = 6$ ways, so there are $10 \cdot 6 = 60$ ways. For category 4, we can choose the pair in 5 ways, the other 2 in $\binom{7}{2} = 21$ ways, and the 4 letters can be arranged in $\binom{4}{2} = 12$ ways for a total of $5 \cdot 21 \cdot 12 = 1,260$ ways. For category 5, we have to choose 4 unique letters from 8, so the number of ways is $8 \cdot 7 \cdot 6 \cdot 5 = 1,680$. The total number of arrangements of 4 letters from the given 15 is then the sum of the arrangements from each category: $1 + 28 + 60 + 1,260 + 1,680 = 3,029$.

Problem 139. In how many ways can an arrangement be made of 3 things chosen from 15 things where 5 are of one type, 4 of another type, 3 of another type, and the last 3 of another type?

Answer. The ways the 3 things can be chosen falls into 3 categories:

1. Three all the same.
2. Two the same, one different.
3. All three different.

For category 1, there are 4 ways to choose all 3 the same. For category 2, there are 4 ways to choose 2 the same, and 3 ways to choose one different, and $\binom{3}{2} = 3$ ways to arrange what was chosen, for a total of $4 \cdot 3 \cdot 3 = 36$ ways. For category 3, there are $4 \cdot 3 \cdot 2 = 24$ ways to choose all 3 different things. The combined number of ways to choose 3 things from the 15 described things is then $4 + 36 + 24 = 64$.

Problem 140. In how many ways can an arrangement be made of 5 things chosen from the 15 things described in the previous problem?

Answer. The ways the 5 things can be chosen falls into 6 categories:

1. Five all the same.

2. Four the same, one different.

3. Three the same, another two the same.

4. Three the same, two different.

5. Two the same, another two the same, one different.

6. Two the same, three different.

For category 1, there is only one way to choose. For category 2, there are 6 ways to choose with 5 ways to arrange for a total of $6 \cdot 5 = 30$ ways. For category 3, there are 12 ways to choose, with $\binom{5}{3,2} = 10$ arrangements, so there are $12 \cdot 10 = 120$ ways. For category 4, there are 12 ways to choose, with $\binom{5}{3,1,1} = 20$ arrangements, so there are $12 \cdot 20 = 240$ ways. For category 5, there are 12 ways to choose, with $\binom{5}{2,2,1} = 30$ arrangements, so there are $12 \cdot 30 = 360$ ways. For category 6, there are 4 ways to choose, with $\binom{5}{2,1,1,1} = 60$ arrangements, so there are $4 \cdot 60 = 240$ ways. The combined number of ways to choose 5 things from the 15 described things is then $1+30+120+240+360+240 = 991$.

Problem 141. How many four letter words can you make with an alphabet of 20 letters if repetitions are allowed?

Answer. There are 20 choices for each of the 4 letters so the number of possible words is: $20 \cdot 20 \cdot 20 \cdot 20 = 20^4 = 160,000$.

Problem 142. How many five letter words can you make with the 5 vowels with repetitions allowed?

Answer. There are 5 choices for each of the 5 letters so the number of possible words is: $5 \cdot 5 \cdot 5 \cdot 5 \cdot 5 = 5^5 = 3,125$.

Problem 143. Using the ten decimal digits 0, 1, 2, 3, 4, 5, 6, 7, 8, 9 how many six digit numbers can you make if leading zeros are not allowed?

Answer. Not allowing leading zeros means that the first digit must be something other than a 0 so there are 9 choices for the first digit. There are 10 choices each for the remaining digits. The number of possible six digit numbers is then: $9 \cdot 10 \cdot 10 \cdot 10 \cdot 10 \cdot 10 = 9 \cdot 10^5 = 900,000$.

Problem 144. There are twelve books for sale. In how many ways can you purchase one or more books?

Answer. If you purchase only one book then you have 12 choices. If you purchase two books then the number of ways of choosing 2 books from 12 is $\binom{12}{2}$. In general if you decide to purchase k books then the number of ways you can choose the k books is $\binom{12}{k}$. The sum of all the possibilities is:

$$\binom{12}{1} + \binom{12}{2} + \cdots + \binom{12}{12}$$

The easy way to do this sum is to remember that these binomial coefficients appear in the expansion:

$$(1+x)^{12} = \binom{12}{0} + \binom{12}{1}x + \binom{12}{2}x^2 \cdots \binom{12}{12}x^{12}$$

If you set $x = 1$ in this equation then the sum on the right is the same as the sum we are trying to calculate except for the $\binom{12}{0} = 1$ term. On the left of the equation we get 2^{12} so subtracting 1 from this gives us the answer: $2^{12} - 1 = 4,095$. In general this question is asking how many subsets of n elements can be formed. If you include the possibility of the empty subset then the answer is 2^n. If you exclude the empty subset then the answer is $2^n - 1$.

Problem 145. How many different weight values can you make by combining five individual weights?

Answer. The answer will depend on the exact values of the five weights. In general they can be combined to give $2^5 - 1 = 31$ weight totals but it may be that not all totals are different. If for example the weights are $2, 4, 6, 8, 10$ then the two different combinations $2 + 8$ and $4 + 6$ both give a value of 10. If the weight values are $1, 2, 4, 8, 16$ then all the possible combinations will be unique and every integer weight value from 1 to 31 will be produced by a combination. In this case the different combinations can be represented as five digit binary numbers. For example 01101 represents the combination $1 + 4 + 8 = 13$. As we know, the number of five digit binary numbers, excluding zero, is $2^5 - 1 = 31$.

Problem 146. A weight w has to be measured on a balance using five counter weights. The counter weights can be placed on either side of the balance. That is they can be put on the same side as the weight to be measured or on the opposite side. How many different w weights can be balanced by the five counter weights.

Answer. In general each counter weight can be put on the side of the weight w, on the opposite side or left off the scale. There are 3 options for each of the five counter weights so there are $3^5 - 1 = 242$ ways they can be used (excluding the case where all 5 are left off the scale). But only half of these combinations will allow a positive weight w to be measured. To see this let w_1, w_2, w_3, w_4, w_5 be the five counter weights then w can be expressed as the sum of the weights placed on the opposite side of the balance minus the sum of the weights placed on the same side. For example if w_1 and w_2 are on the same side as w, and w_5 is on the opposite side then we must have $w + w_1 + w_2 = w_5$ or $w = w_5 - w_1 - w_2$. You can now multiply this last equation by -1 which amounts to switching the counter weights from one side to the other. But then the measured weight w is negative. Therefor for every combination that allows a positive weight to be measured there will be a combination resulting in a negative measured weight. If these latter combinations are excluded then there are only $\frac{242}{2} = 121$ valid combinations. Whether all these combinations measure a unique weight

w will depend on the exact values of the counter
weights. The weight set $1, 3, 9, 27, 81$ will allow
121 unique weights to be measured. You can see
this by noting that there is a one to one correspondence between combinations of the counter
weights and 5 digit trinary numbers. Each combination of counter weights can be written as:

$$w = c_1 w_1 + c_2 w_2 + c_3 w_3 + c_4 w_4 + c_5 w_5 \quad (69)$$

where $c_i = 1$ if w_i is placed opposite to w, $c_i = -1$
if w_i is placed on the same side as w and $c_i = 0$ if
w_i is left off the scale. A 5 digit trinary number
on the other hand can be written as:

$$t_0 + t_1 3 + t_2 3^2 + t_3 3^3 + t_4 3^4 \quad (70)$$

where each trinary digit can be 0, 1, or 2. If
$w_i = 3^i$ and $c_i = t_{i-1} - 1$ then there is a one to
one correspondence between the unique values of
both equations.

Problem 147. In how many ways can two booksellers
split an inventory of 200 copies of Tarzan, 250
copies of Robinson Crusoe, 150 copies of Don
Quixote, and 100 copies of Marco Polo?

Answer. One of the booksellers can take anywhere
from 0 to 200 copies of Tarzan with the other
taking the rest so there are 201 ways they can

split Tarzan. Similarly there are 251 ways to split Robinson Crusoe, 151 ways to split Don Quixote, and 101 ways to split Marco Polo. The total number of ways they can split the books is then: $201 \cdot 251 \cdot 151 \cdot 101 = 769,428,201$. In one of these ways the first bookseller gets none of the books and in another one of the ways the second bookseller gets none of the books. If you exclude these two possibilities then the number of ways they can split the books is equal to $769,428,201 - 2 = 769,428,199$.

Problem 148. What is the total number of selections you can make from the letters *ned needs nineteen nets*?

Answer. There are 20 letters, with 6 *n*'s, 7 *e*'s, 2 *d*'s, 2 *s*'s, 1 *i*, and 2 *t*'s. From the *n*'s we can choose none or up to 6, making 7 choices. For the *e*'s we can choose none or up to 7, making 8 choices, and similarly for the remaining letters. So the total number of selections we can make is $7 \cdot 8 \cdot 3 \cdot 3 \cdot 2 \cdot 3 = 3,024$. One of these selections includes the case of not choosing any letters at all, so the total is $3,023$ selections.

Problem 149. What is the total number of selections you can make from the letters *daddy did a nice deed*?

Answer. There are 17 letters, with 7 d's, 2 a's, 1 y, 2 i's, 1 n, 1 c, and 3 e's. Similarly to the last question, the total number of selections we can make is $8 \cdot 3 \cdot 2 \cdot 3 \cdot 2 \cdot 2 \cdot 4 = 2,304$. One of these selections includes the case of not choosing any letters at all, so the total is $2,303$ selections.

Problem 150. From the letters in the last question, how many selections of 3 letters can be made?

Answer. The 3 chosen letters can fall into any of 3 categories:

1. all 3 letters identical
2. 2 letters identical, one different
3. all 3 letters different

For the first category there are 2 ways. For the second category there are 4 ways to choose 2 identical letters, and 6 ways to choose the different letter, which makes $4 \cdot 6 = 24$ ways. For the third category there are $\binom{7}{3} = \frac{7!}{3! \cdot 4!} = 35$ ways to choose 3 different letters. The total number of selections that can be made is then $2 + 24 + 35 = 61$.

Problem 151. From the last question, how many arrangements of 3 letters can be made?

Answer. For category (1) there are 2 arrangements of the 2 ways. For category (2) there are $24 \cdot \binom{3}{2,1} =$

72 arrangements of the 24 ways. And for category
(3) there are $35 \cdot 3! = 210$ arrangements of the 35
ways. The total number of arrangements is then
$2 + 72 + 210 = 284$.

Problem 152. Show that there are 8 subsets of size
3, from the letters of *veneer*, and 16 subsets of
size 4 from the letters of *veneered*.

Answer. *veneer* has 6 letters with 3 *e*'s, and the rest
single occurrences. Subsets of size 3 fall into 3
categories:

1. All 3 letters are the same. There is only one
   way for this to occur.

2. Just 2 letters are the same, one is different.
   There are 3 ways for this to happen.

3. All 3 letters are different. There are $\binom{4}{3} = \frac{4!}{3!} = 4$ ways for this.

The total number of subsets of size 3 is then $1 + 3 + 4 = 8$. *veneered* has 8 letters with 4 *e*'s, and
the rest single occurrences. There are 4 categories
for subsets of size 4:

1. All 4 letters are the same. There is only one
   way for this to occur.

2. Just 3 letters are the same, one is different.
   There are $\binom{4}{1} = 4$ ways for this to happen.

3. Just 2 letters are the same, 2 are different. There are $\binom{4}{2} = 6$ ways for this to happen.

4. All 4 letters are different. There are $\binom{5}{4} = \frac{5!}{4!} = 5$ ways for this to happen.

The total number of subsets of size 4 is then $1 + 4 + 6 + 5 = 16$.

Problem 153. How many subsets of size 3 are there in the letters *wedded*?

Answer. *wedded* has 6 letters with 2 *e*'s, 3 *d*'s, and one *w*. Subsets of size 3 fall into 3 categories:

1. All 3 letters are the same. There is only one way for this to occur.

2. Just 2 letters are the same, one is different. There are $2 \cdot 2 = 4$ ways for this to occur.

3. All 3 letters are different. There is $\binom{3}{3} = 1$ way for this to occur.

The total number of subsets of size 3 is then $1 + 4 + 1 = 6$.

Problem 154. How many subsets of size 4 are there in the letters *redeemed*?

Answer. *redeemed* has 8 letters with 4 *e*'s, 2 *d*'s, and one occurrence each of *r* and *m*. Subsets of size 4 fall into 5 categories:

1. All 4 letters are the same. There is only one way for this to occur.

2. Just 3 letters are the same, one is different. There are 3 ways for this to occur.

3. Only 2 letters are the same, 2 are different. There are $2 \cdot \binom{3}{2} = 6$ ways for this to occur.

4. 2 letters are the same, the other 2 are the same. There is one way for this to occur.

5. All 4 letters are different. There is $\binom{4}{4} = 1$ way for this to occur.

The total number of subsets of size 4 is then $1 + 3 + 6 + 1 + 1 = 12$.

Problem 155. How many subsets of size 5 are there in the letters *ever esteemed*?

Answer. *ever esteemed* has 12 letters with 6 *e*'s, and the rest single occurrences. Subsets of size 5 fall into 5 categories:

1. All 5 letters are the same. There is only one way for this to occur.

2. Just 4 letters are the same, one is different. There are 6 ways for this to occur.

3. Only 3 letters are the same, 2 are different. There are $\binom{6}{2} = 15$ ways for this to occur.

4. 2 letters are the same, the other 3 are different. There are $\binom{6}{3} = 20$ ways for this to occur.

5. All 5 letters are different. There are $\binom{7}{5} = 21$ ways for this to occur.

The total number of subsets of size 5 is then $1 + 6 + 15 + 20 + 21 = 63$.

Problem 156. How many ways can 3 people distribute the letters of *ever esteemed* among each other?

Answer. *ever esteemed* has 12 letters with 6 *e*'s, and the rest single occurrences. Distributing the 6 *e*'s is equivalent to the question "How many ways can you place 6 identical balls into 3 distinct bins with no restrictions?". The answer is $\binom{6+3-1}{3-1} = \binom{8}{2} = 28$ ways. Distributing the remaining 6 unique letters is equivalent to the question "How many ways can you place 6 distinct balls into 3 distinct bins with no restrictions?". The answer is $3^6 = 729$ ways. The total number of ways is then $28 \cdot 729 = 20,412$. But this includes the cases where either one or 2 people end up with nothing. There are 3 ways for 2 people to end up with nothing. For only one person to end up with nothing, the other 2 people must get something, and that can be done in $\binom{6+2-1}{2-1} \cdot 2^6 - 2 = 7 \cdot 64 - 2 = 446$ ways, where the 2 is subtracted for the cases of either of the 2 people getting everything. Since there are 3 people who can end up with nothing, we must multiply this result by 3 to get $446 \cdot 3 = 1,338$. So removing the cases where someone ends up with nothing, we get $20,412 - 3 - 1,338 = 19,071$ ways.

Problem 157. For the previous question, how many ways can the letters be divided so that each person gets 4?

Answer. This is equivalent to the question "If you have 12 balls, with 6 being distinct and 6 identical, how many ways can you put 4 balls each into 3 distinct urns?". We'll look at this problem from the standpoint of putting the distinct balls into the urns first, then topping off with identical balls to end with 4 balls in each urn. If there were no restrictions on the number of distinct balls placed into each urn, then the answer would be $3^6 = 729$. Now we just need to subtract the number of ways that any urn can get 6 or 5 balls. All 6 balls can end up in a certain urn in 1 way. 5 balls can end up in a certain urn in $\binom{6}{5} = 6$ ways, with 2 ways to place the remaining ball, making $6 \cdot 2 = 12$ ways. Because we have 3 urns, the number of ways to get 6 or 5 balls in an urn is then $3 \cdot (1 + 12) = 39$ ways. The total number of ways is then $729 - 39 = 690$. Filling each urn up to the required 4 with identical balls can only be done in one way, so the result is still 690.

Problem 158. Albert has the six letters *esteem*, Bob has the six letters *feeble*, and Carter has the six letters *veneer*. In how many ways can the letters be redistributed so each of them still has six letters?

Answer. There are a total of 18 letters with 9 of them being e's and the other 9 all different. First distribute the different letters so that no one has more than 6 and then distribute the e's to give everyone exactly 6 letters. The 9 different letters can be distributed in a total of $3^9 = 19,683$ ways but in some of these ways a person gets more than 6 letters. Albert can get 9 letters in only one way. He can get 8 letters in 9 ways with Bob and Carter getting the remaining letter in 2 ways. He can get 7 letters in $\binom{9}{7} = 36$ ways and the other two get the remaining 2 letters in $2^2 = 4$ ways. The number of ways Albert gets more than 6 letters is then $1 + 9 \cdot 2 + 36 \cdot 4 = 163$. The same applies to Bob and Carter. This means there are $3 \cdot 163 = 489$ ways in which someone gets more than 6 of the different letters. The different letters can therefor be distributed in $19,683 - 489 = 19,194$ correct ways. For each of these ways there is only one way to distribute the e's so that each person has exactly 6 letters. The answer to the question is then $19,194$.

Problem 159. How many ways can you arrange the letters in the word $falsity$ so that the consonants f, l, s, t and the vowels a, i, y keep the same order?

Answer. You can choose 4 of the 7 letters to be consonants and the remaining 3 to be vowels in $\binom{7}{4} = 35$ ways. For each of these ways there is only one way to place the consonants and vowels to keep the same order so the answer is 35.

Problem 160. How many ways can you arrange the letters in the word $affection$ so that the vowels keep their order and the two f's are separated?

Answer. There are a total of 9 letters with 5 consonants and 4 vowels. The number of ways places for the consonants can be selected is $\binom{9}{5}$. Assuming the 2 f's are distinguishable, there are 5! ways to arrange the consonants once the places have been chosen. The distinguishability assumption for the f's can be removed by dividing by 2. For every way of arranging the consonants there is only one way to arrange the vowels to keep their order. So the number of ways to arrange the letters without restriction on keeping the f's apart is $\binom{9}{5}\frac{5!}{2} = \frac{9!}{4!2} = 7,560$. To find the number of these arrangements where the f's are together, treat ff as a new consonant. Now there are 8 letters with 4 consonants and 4 vowels. The number of ways they can be arranged is $\binom{8}{4}4! = \frac{8!}{4!} = 1,680$. Subtract this from the previous number to get the number of arrangements with the f's separated: $7,560 - 1,680 = 5,880$.

Problem 161. How many ways can you arrange the letters of $kaffeekanne$ so that each arrangement alternates between consonant and vowel?

Answer. There are 6 consonants $kkffnn$ that can be arranged in $\binom{6}{2,2,2} = \frac{6!}{2^3} = 90$ ways. There are 5 vowels $aaeee$ that can be arranged in $\binom{5}{2,3} = \frac{5!}{12} = 10$ ways. For each arrangement of consonants, pick an arrangement of vowels to place between the consonants. The number of ways to do this is then $90 \cdot 10 = 900$.

Problem 162. How many arrangements of the word $delete$ keep the order of the consonants?

Answer. $delete$ has 6 letters, with 3 consonants dlt, and 3 vowels which are all e's. This problem can be thought of in terms of 6 urns, with 3 of the urns to be chosen to place the consonants in their fixed order. 3 urns out of 6 can be chosen in

$\binom{6}{3} = 20$ ways. The remaining 3 urns are filled each with one e, which can be done in only one way. So the total number of arrangements is 20.

Problem 163. For the previous question, how many arrangements are there if we add the restriction of not having any 2 e's together?

Answer. There are 4 ways: *delete*, *edelet*, *edlete*, and *edelte*.

Problem 164. How many ways can the letters of *delirious* be arranged keeping the vowels and consonants in their original order?

Answer. *delirious* has 9 letters with 4 consonants and 5 vowels. Places for the consonants can be chosen in $\binom{9}{4} = 126$ ways. Vowels and consonants must be kept in order so there is only one way to arrange them for each of these ways. The answer is then 126.

Problem 165. For the previous question, what is the number of ways if we exclude arrangements with 2 i's side by side?

Answer. Treat the 2 i's as a single vowel, so there are 8 letters with 4 consonants and 4 vowels. The consonant places can be chosen in $\binom{8}{4} = 70$ ways.

In each of these ways the consonants and vowels can only be filled in one way. So there are 70 arrangements where the 2 i's appear together. Subtract this from the previous answer to get $126 - 70 = 56$ ways to arrange things without the 2 i's together.

Problem 166. How many ways can the letters $fulfil$ be arranged so that no 2 consecutive letters are the same?

Answer. $fulfil$ has 6 letters, with 2 f's, 1 u, 2 l's, and 1 i. Without restriction, the letters can be arranged in $\binom{6}{2,2} = \frac{6!}{4} = 180$ ways. Treating ff as a single letter, the number of arrangements are $\binom{5}{2,1,1,1} = \frac{5!}{2} = 60$. Likewise, treating ll as a single letter, the number of arrangements are $\binom{5}{2,1,1,1} = \frac{5!}{2} = 60$. Treating ff and ll as single letters, the number of arrangements are $\binom{4}{1,1,1,1} = 4! = 24$. The number of arrangements where ff, ll, or both appear is then $60 + 60 - 24 = 96$. So the number of arrangements where neither ff or ll appear is $180 - 96 = 84$.

Problem 167. How many ways can the letters $murmur$ be arranged so that no 2 consecutive letters are alike?

Answer. $murmur$ has 6 letters, with 2 m's, 2 u's, and 2 r's. The number of unrestricted arrangements is $\binom{6}{2,2,2} = \frac{6!}{8} = 90$. Let

A = set of arrangements with 2 consec. m's
B = set of arrangements with 2 consec. u's
C = set of arrangements with 2 consec. r's

The number of members of A, B, and C is then

$$|A| = |B| = |C| = \binom{5}{2,2,1} = \frac{120}{4} = 30$$

The number of common members between the sets is

$$|A \cap B| = |A \cap C| = |B \cap C|$$
$$= \binom{4}{2} = 12$$
$$|A \cap B \cap C| = 3! = 6$$

$A \cup B \cup C =$
set of arrangements with at least 1 double letter

$$\begin{aligned}
|A \cup B \cup C| = {} & |A| + |B| + |C| \\
& - |A \cap B| - |A \cap C| \\
& - |B \cap C| \\
& + |A \cap B \cap C| \\
= {} & 3 \cdot 30 - 3 \cdot 12 + 6 \\
= {} & 3 \cdot 18 + 6 \\
= {} & 54 + 6 = 60
\end{aligned}$$

Subtracting this from the unrestricted arrangements gives $90 - 60 = 30$.

Problem 168. How many ways can you select 4 letters from *murmur* and how many arrangements are possible?

Answer. There are 3 letters with multiplicities of 2 each. When selecting 4 letters from these 6, there will be either 2 letters with multiplicity 2 each, or one with multiplicity 2 and the other 2 different. 2 with multiplicity 2 can be selected in $\binom{3}{2} = 3$ ways. One with multiplicity 2, and the other 2 different can be selected in $\binom{3}{1}\binom{2}{2} = 3$ ways. The number of selections is then $3 + 3 = 6$. There are $\binom{4}{2,2} = 6$ ways to arrange the $(2, 2)$ multiplicities, and $\binom{4}{2,1,1} = 12$ ways to arrange the $(2, 1, 1)$ multiplicities, so the number of arrangements is $3 \cdot 6 + 3 \cdot 12 = 18 + 36 = 54$.

Problem 169. How many 4 letter words can you make from the letters *fulfil*?

Answer. *fulfil* has 4 unique letters (f, l, i, u) with multiplicities $(2, 2, 1, 1)$ for a total of 6 letters. A selection of 4 letters can be made with multiplicities $(2, 2)$ or $(2, 1, 1)$, or $(1, 1, 1, 1)$. The $(2, 2)$ selection can be made in one way and the resulting letters can be arranged in $\binom{4}{2,2} = \frac{4!}{2^2} = 6$ ways.

The $(2, 1, 1)$ selection can be made in $\binom{2}{1}\binom{3}{2} = 2 \cdot 3 = 6$ ways and the resulting letters arranged in $\binom{4}{2} = 12$ ways. The $(1, 1, 1, 1)$ selection can be made in one way, and the resulting letters arranged in $4! = 24$ ways. The total number of words is then $1 \cdot 6 + 6 \cdot 12 + 1 \cdot 24 = 6 + 72 + 24 = 102$.

Problem 170. How many 5 letter words can you make from the letters *pallmall*?

Answer. There are 4 unique letters (l, a, m, p) with multiplicities $(4, 2, 1, 1)$ for a total of 8 letters. The multiplicities of a selection of 5 letters is shown in table 6. The total number of 5 letter

Mult	Selections	Arrangements	Select·Arrang
(4,1)	3	$\frac{5!}{4!} = 5$	15
(3,2)	1	$\frac{5!}{3! \cdot 2!} = 10$	10
(3,1,1)	3	$\frac{5!}{3!} = 20$	60
(2,2,1)	2	$\frac{5!}{2! \cdot 2!} = 30$	60
(2,1,1,1)	2	$\frac{5!}{2!} = 60$	120

Table 6: Multiplicities for Problem 170.

words is then $15 + 10 + 60 + 60 + 120 = 265$.

Problem 171. How many 4 letter words can you make from the letters *kaffeekanne* excluding words with 3 *e*'s in a row?

Answer. There are 5 unique letters (e, k, a, f, n) with multiplicities $(3, 2, 2, 2, 2)$ for a total of 11 letters. The multiplicities for a selection of 4 letters are shown in table 7. Without restrictions the

Mult	Selections	Arrangements	Select·Arrang
(3,1)	4	$\frac{4!}{3!} = 4$	16
(2,2)	$\binom{5}{2} = 10$	$\frac{4!}{2! \cdot 2!} = 6$	60
(2,1,1)	$\binom{5}{1}\binom{4}{2} = 30$	$\frac{4!}{2!} = 12$	360
(1,1,1,1)	$\binom{5}{4} = 5$	$4! = 24$	120

Table 7: Multiplicities for Problem 171.

number of words is $16 + 60 + 360 + 120 = 556$. There are 4 ways to select the 3 e's and one of the other letters. In each of these ways treat the 3 e's as a single letter then it can be arranged in 2 ways with the other letter. The number of words with 3 e's together is then $4 \cdot 2 = 8$. Subtracting this from the unrestricted word count gives $556 - 8 = 548$ words.

Problem 172. How many 6 letter words can you make from the letters *nineteen tennis nets*?

Answer. There are 5 unique letters (n, e, t, i, s) with multiplicities $(6, 5, 3, 2, 2)$ for a total of 18 letters. When selecting 6 letters, the possible multiplicities, ways to select, and arrangements are in table 8. The total number of 6 letter words is then got-

Mult.	Sel.	Arr.	Sel.·Arr.
(6)	1	1	1
$(5,1)$	$\binom{2}{1}\binom{4}{1} = 8$	$\frac{6!}{5!} = 6$	48
$(4,2)$	$\binom{2}{1}\binom{4}{1} = 8$	$\frac{6!}{4! \cdot 2!} = 15$	120
$(4,1,1)$	$\binom{2}{1}\binom{4}{2} = 12$	$\frac{6!}{4!} = 30$	360
$(3,3)$	$\binom{3}{2} = 3$	$\frac{6!}{3! \cdot 3!} = 20$	60
$(3,2,1)$	$\binom{3}{1}\binom{4}{1}\binom{3}{1} = 36$	$\frac{6!}{3! \cdot 2!} = 60$	2160
$(3,1,1,1)$	$\binom{3}{1}\binom{4}{3} = 12$	$\frac{6!}{3!} = 120$	1440
$(2,2,2)$	$\binom{5}{3} = 10$	$\frac{6!}{(2!)^3} = 90$	900
$(2,2,1,1)$	$\binom{5}{2}\binom{3}{2} = 30$	$\frac{6!}{(2!)^2} = 180$	5400
$(2,1,1,1,1)$	$\binom{5}{1}\binom{4}{4} = 5$	$\frac{6!}{2!} = 360$	1800

Table 8: Multiplicities for Problem 172.

ten by summing up the right column of the table, which is 12,289.

Problem 173. How many 6 letter words can you make with the letters *littlepipe* if no letter can follow itself?

Answer. There are 5 unique letters (l, i, t, e, p) with multiplicities $(2, 2, 2, 2, 2)$ for a total of 10 letters. First find the number of 6 letter words with no restrictions. The possible multiplicities, ways to select, and arrangements are in table 9. The number of unrestricted words is then gotten by summing up the right column of table 9, giving 8,100. With the $(2, 2, 2)$ multiplicity, we have 2

Mult.	Sel.	Arr.	Sel.·Arr.
(2,2,2)	$\binom{5}{3} = 10$	$\frac{6!}{(2!)^3} = 90$	900
(2,2,1,1)	$\binom{5}{2}\binom{3}{2} = 30$	$\frac{6!}{(2!)^2} = 180$	5400
(2,1,1,1,1)	$\binom{5}{1} = 5$	$\frac{6!}{2!} = 360$	1800

Table 9: Multiplicities for Problem 173.

copies each of 3 unique letters. Let A, B, and C be the sets containing words with the first, second, and third of the unique letters appearing double. We want to find

$$\begin{aligned} |A \cup B \cup C| =& |A| + |B| + |C| \\ &- |A \cap B| - |A \cap C| \\ &- |B \cap C| \\ &+ |A \cap B \cap C| \end{aligned}$$

where

$$|A| = |B| = |C| = \frac{5!}{2! \cdot 2!} = 30$$
$$|A \cap B| = |A \cap C| = |B \cap C|$$
$$= \frac{4!}{2!} = 12$$
$$|A \cap B \cap C| = 3! = 6$$

so that

$$|A \cup B \cup C| = 3 \cdot 30 - 3 \cdot 12 + 6 = 60$$

There are 10 ways to have a $(2, 2, 2)$ multiplicity so there are a total of $60 \cdot 10 = 600$ restricted words for this multiplicity. With the $(2, 2, 1, 1)$ multiplicity there are 2 copies of 2 of the letters. Let A and B be the sets containing words where the first and second of these letters appear double. We want to find

$$|A \cup B| = |A| + |B| - |A \cap B|$$

where

$$|A| = |B| = \frac{5!}{2!} = 60$$
$$|A \cap B| = 4! = 24$$

so that

$$|A \cup B| = 2 \cdot 60 - 24 = 96$$

There are 30 ways to have a $(2, 2, 1, 1)$ multiplicity so there are $96 \cdot 30 = 2,880$ restricted words. For the $(2, 1, 1, 1, 1)$ multiplicity there are $5! = 120$ words with a double letter. There are 5 ways to get the multiplicity, so there are a total of $120 \cdot 5 = 600$ restricted words. The total number of restricted words is $600 + 2,880 + 600 = 4,080$. The number of acceptable words is then $8,100 - 4,080 = 4,020$.

Problem 174. How many 5 letter words can you make with the letters *murmurer* so that the 3 *r*'s do not appear together?

Answer. There are 4 unique letters (r, m, u, e) with multiplicities $(3, 2, 2, 1)$ for a total of 8 letters. First find the number of 5 letter words with no restrictions. The possible letter multiplicities, ways to select and arrangements are in table 10. With

Mult.	Sel.	Arr.	Sel.·Arr.
(3,2)	2	$\frac{5!}{3!\cdot 2!} = 10$	20
(3,1,1)	3	$\frac{5!}{3!} = 20$	60
(2,2,1)	$\binom{3}{2}\binom{2}{1} = 6$	$\frac{5!}{(2!)^2} = 30$	180
(2,1,1,1)	$\binom{3}{1}\binom{3}{3} = 3$	$\frac{5!}{2!} = 60$	180

Table 10: Multiplicities for Problem 174.

no restrictions, the number of 5 letter words is the sum of the right column, which is 440. The $(3, 2)$ multiplicity will have 3 *r*'s. Treat them as a single letter so that the number of words is $\binom{3}{2} = \frac{3!}{2!} = 3$. There are 2 ways to get this multiplicity, so the number of restricted words is $3 \cdot 2 = 6$. The $(3, 1, 1)$ multiplicity also has 3 *r*'s. Treating them as a single letter, the number of words is $3! = 6$. There are 3 ways to get the multiplicity, so the number of restricted words is $6 \cdot 3 = 18$. The total number of restricted words is $6 + 18 = 24$. The number of acceptable words is then $440 - 24 = 416$.

Problem 175. How many ways can you arrange the letters *quisquis* so that no letter follows itself?

Answer. There are 2 copies each of the letters (q, u, i, s). The number of unrestricted arrangements is

$$\frac{8!}{(2!)^4} = 2,520$$

Define
$S_1 =$ number of words where 1 particular letter appears doubled.
$S_2 =$ number of words where 2 particular letters appear doubled. Likewise for S_3 and S_4. The number of restricted words is then

$$\binom{4}{1} S_1 - \binom{4}{2} S_2 + \binom{4}{3} S_3 - \binom{4}{4} S_4$$
$$= 4S_1 - 6S_2 + 4S_3 - S_4$$

$$S_1 = \frac{7!}{(2!)^3} = \frac{7!}{8} = 630$$
$$S_2 = \frac{6!}{(2!)^2} = \frac{6!}{4} = 180$$
$$S_3 = \frac{5!}{2!} = 60$$
$$S_4 = 4! = 24$$

The total number of restricted words is $4 \cdot 630 - 6 \cdot 180 + 4 \cdot 60 - 24 = 1,656$. So the number of acceptable words is $2,520 - 1,656 = 864$.

Problem 176. How many ways can you arrange the letters *feminine* so that no letter follows itself?

Answer. The unique letters are (e, i, n, m, f) with multiplicities $(2, 2, 2, 1, 1)$ for a total of 8 letters. The number of unrestricted arrangements is

$$\frac{8!}{(2!)^3} = \frac{8!}{8} = 7! = 5,040$$

Define S_i = number of words where i particular letters appear doubled. The number of restricted words is then

$$\binom{3}{1} S_1 - \binom{3}{2} S_2 + \binom{3}{3} S_3 = 3S_1 - 3S_2 + S_3$$

$$S_1 = \frac{7!}{2!2!} = 1,260$$
$$S_2 = \frac{6!}{2!} = 360$$
$$S_3 = 5! = 120$$

The total number of restricted words is $3 \cdot 1,260 - 3 \cdot 360 + 120 = 2,820$. So the number of acceptable words is $5,040 - 2,820 = 2,220$.

Problem 177. How many ways can you arrange the letters *muhammadan* so 3 identical letters do not appear together?

Answer. Unique letters and multiplicities are (a, m, d, h, n, u) $(3, 3, 1, 1, 1, 1)$ for a total of 10 letters. The number of unrestricted arrangements is

$$\frac{10!}{3!3!} = 100,800$$

Let S_i be the number of words where i particular letters appear triple. The number of restricted words is then

$$\binom{2}{1} S_1 - \binom{2}{2} S_2 = 2S_1 - S_2$$

$$S_1 = \frac{8!}{3!} = 6,720$$
$$S_2 = 6! = 720$$

So $2S_1 - S_2 = 12,720$ is the total number of restricted words. The number of acceptable words is then $100,800 - 12,720 = 88,080$.

Problem 178. How many ways can you arrange the letters *muhammadan* so that 2 identical letters do not appear together?

Answer. The unique letters and multiplicities are (a, m, d, h, n, u) $(3, 3, 1, 1, 1, 1)$. Start by looking at the arrangements without the letter a. There are $\binom{7}{3,1,1,1,1} = \frac{7!}{3!} = 840$ such arrangements of which $5! = 120$ have 3 m's in a row (treat 3 m's in a row as one symbol). Take 2 of the m's to be a new letter, then there are 6 letters that can be arranged in $6! = 720$ ways of which $5!$ have 3 m's in a row. The number of arrangements with only 2 m's in a row is then $6! - 5! = 600$. The number of arrangements with all m's single is $840 - 120 - 600 = 120$. There are 3 kinds of arrangements with no a's.

1. 120 arrangements with only mmm's

2. 600 arrangements with only mm's

3. 120 arrangements with only m's

Now combine the a's with these arrangements. For the mmm arrangements, 2 of the a's must be used to separate the m's. There are 6 possible positions for the remaining a. For the mm arrangements, one of the a's must be used to separate the 2 m's. There are 7 possible positions for the remaining 2 a's. For the arrangements with only single m's, there are 8 possible positions for the 3 a's. The number of arrangements of all the letters is therefore

$$\binom{6}{1}120 + \binom{7}{2}600 + \binom{8}{3}120$$
$$= 720 + 21 \cdot 600 + 56 \cdot 120$$
$$= 20,040$$

Problem 179. Given 20 consecutive numbers, how many ways can you select 2 of them so that they sum to an odd number?

Answer. For 2 numbers to sum to an odd number, one of them must be odd and the other even. In any set of 20 consecutive numbers, 10 will be even and 10 will be odd, so there are $10 \cdot 10 = 100$ ways to make the selection.

Problem 180. Given 30 consecutive numbers, how many ways can you select 3 to get an even sum?

Answer. Let e represent an even number, and o an odd number, then the possible combinations are:

$$e + e + e = e$$
$$e + e + o = o$$
$$e + o + o = e$$
$$o + o + o = o$$

So the multiplicities of (even,odd) numbers must be $(3,0)$ or $(1,2)$ to get an even sum. There are

15 even and 15 odd numbers so the number of ways to select is

$$\binom{15}{3}\binom{15}{0} + \binom{15}{1}\binom{15}{2} = 2,030$$

Problem 181. How many ways can you collect 20 coins composed of pennies, nickels, and dimes?

Answer. Think of this in terms of distributing 20 identical balls into 3 distinct urns. The 3 urns represent pennies, nickels, and dimes, and a particular distribution of the balls corresponds to a particular collection of coins. The number of ways to distribute n balls into k urns is $\binom{n+k-1}{k-1}$. With $n = 20$ and $k = 3$ this is $\binom{22}{2} = 231$.

Problem 182. A person has 3 coins and she asks you to guess what they are. Given that the coins could be pennies, nickels, dimes, quarters, half dollars, or dollars, how many guesses do you have to make to guarantee a correct answer?

Answer. The number of possible guesses is equal to the number of ways you can collect 3 coins of 6 possible types. Like the previous problem, this is equal to the number of ways to distribute 3 balls into 6 urns or $\binom{8}{5} = 56$. So if you are very unlucky, then you will have to make 56 guesses before the last one is finally correct.

Problem 183. How many 5 digit decimal numbers are there (no leading zeros allowed)? In how many of them is every digit odd? In how many is every digit even? In how many are there no digits less than 6? In how many are there no digits greater than 3? How many contain all the digits 1,2,3,4,5? How many contain all the digits 0,2,4,6,8?

Answer. There are $9 \cdot 10 \cdot 10 \cdot 10 \cdot 10 = 90,000$ 5 digit numbers. There are $5 \cdot 5 \cdot 5 \cdot 5 \cdot 5 = 3,125$ numbers with all odd digits. There are $4 \cdot 5 \cdot 5 \cdot 5 \cdot 5 = 2,500$ with all even digits. There are 4 digits ≥ 6, so there are $4^5 = 1,024$ numbers. There are 4 digits ≤ 3, but the first digit cannot be 0 so there are $3 \cdot 4^4 = 768$ numbers. There are $5! = 120$ numbers containing all the digits 1,2,3,4,5. Using all the digits 0,2,4,6,8 there are 4 choices for the first digit and the remaining digits can be arranged in 4! ways, so there are $4 \cdot 4! = 96$ numbers.

Problem 184. 2 dice with faces numbered 0, 1, 3, 7, 15, 31 are thrown. How many different sums are possible?

Answer. This is equal to the number of ways to distribute 2 identical balls into 6 distinct urns

$$\binom{2+5}{5} = \binom{7}{5} = \frac{6 \cdot 7}{2} = 21$$

Problem 185. 3 dice with faces 1, 4, 13, 40, 121, 364 are thrown. How many different sums are possible?

Answer. This is equal to the number of ways to distribute 3 identical balls into 6 distinct urns

$$\binom{3+5}{5} = \binom{8}{5} = \frac{6 \cdot 7 \cdot 8}{6} = 56$$

Problem 186. The post office sells 10 kinds of stamps. How many ways can a person buy 12 stamps? How many ways can a person buy 8 stamps? How many ways can a person buy 8 different stamps?

Answer. For buying 12 stamps, it's like putting 12 identical balls into 10 distinct urns, which can be done in

$$\binom{12+9}{9} = \binom{21}{9} = 293,930$$

ways. Similarly, buying 8 stamps can be done in

$$\binom{8+9}{9} = \binom{17}{9} = 24,310$$

ways. A person can buy 8 different stamps in the same number of ways that 8 things can be selected from 10 things, which is

$$\binom{10}{8} = \frac{9 \cdot 10}{2} = 45$$

ways.

Problem 187. How many ways can you deal 4 cards to each of 13 players so that each gets one card of each suit? How many ways are there if one person gets one card of each suit and the 12 others each get 4 cards of a single suit?

Answer. Assume a standard deck of 52 cards with 4 suits of 13 cards each. The 13 cards of a suit must be distributed one to a player. Each distribution amounts to a permutation of the cards, and there are 13! permutations per suit. There are 4 suits so the number of ways to distribute the cards is $(13!)^4$. For the second part of the question, the single person can be chosen in 13 ways and then receive the 4 cards in 13^4 ways, so giving one person one card from each suit can be done in 13^5 ways. Now there are 4 suits of 12 cards each left to distribute to the remaining 12 players. Each suit can be divided into 3 groups of 4 cards in $\frac{12!}{(4!)^3}$ ways. This includes permutations of the groups, and we only want combinations, so we must divide this by 3! to get $\frac{12!}{(4!)^3 3!}$ ways. So all 4 suits can be divided into 12 groups of 4 cards (all of the same suit) in $\left(\frac{12!}{(4!)^3 3!}\right)^4$ ways. These 12 groups can be distributed to the 12 players in 12! ways. The number of ways to distribute the cards is then

$$13^5 \frac{(12!)^4}{(4!)^{12}(3!)^4} 12! = \frac{(13!)^5}{(4!)^{12}(3!)^4} = \frac{(13!)^5}{2^{40}3^{16}}$$
$$= 197,816,120,269,284,346,830,000,000,000$$
$$\approx 1.98 \times 10^{29}$$

Problem 188. How many ways can you deal 52 cards to 4 players so that each player has 3 cards each of 3 suits and 4 cards of the remaining suit?

Answer. Assigning the players the suit of which they have 4 cards can be done in 4! ways. Each suit can then be distributed in $\frac{13!}{4! \cdot 3! \cdot 3! \cdot 3!}$ ways and all the cards can be distributed in

$$4! \left(\frac{13!}{4!(3!)^3} \right)^4 = 49,965,764,397,515,366,400,000,000$$
$$\approx 4.997 \times 10^{25}$$

ways.

Problem 189. How many ways can 18 unique things be distributed to 5 people so that 4 get 4 things and 1 gets 2? How many ways can they be distributed if 3 people get 4 things and 2 get 3 things?

Answer. The set of 18 can be divided into 4 sets of 4 and one set of 2 in $\frac{18!}{4! \cdot 4! \cdot 4! \cdot 4! \cdot 2!}$ ways. For each such division, there are $\binom{5}{4} = 5$ ways the sets can be distributed. The total number of ways to distribute the 18 things is then

$$\frac{5 \cdot 18!}{(4!)^4 2!} = \frac{5 \cdot 17!}{2^{12} \cdot 3^2}$$
$$= 48,243,195,000$$

The set of 18 can be divided into 3 sets of 4 and 2 sets of 3 in $\frac{18!}{4! \cdot 4! \cdot 4! \cdot 3! \cdot 3!}$ ways. For each division, there are $\binom{5}{3} = 10$ ways the sets can be distributed. The total number of ways to distribute the 18 things is then

$$\frac{10 \cdot 18!}{(4!)^3 (3!)^2} = \frac{5 \cdot 17!}{2^9 \cdot 3^3}$$
$$= 128,648,520,000$$

.

Problem 190. There are 14 kinds of things with 2 of each kind. How many different selections are possible?

Answer. For each of the 14 kinds you have 3 choices. You can select 0, 1, or 2 of that kind. The number of selections is then $3^{14} = 4,782,969$. Note that this includes the case of selecting none at all.

Problem 191. With 20 kinds of things and 9 of each kind, how many different selections can you make?

Answer. Analogous to the previous question there are 10 choices for each thing, so the total number of selections is 10^{20}. Again, this includes the case of selecting none at all.

Problem 192. Bagatelle is a game similar to billiards. There are 9 balls and 9 different holes. Each hole can hold only one ball, and the object is to get the balls in the holes. In a given state of play, some set of holes will be occupied by balls. If there are 8 white balls and one black ball, how many states can the game be in?

Answer. First consider states where the black ball occupies one of the holes. There are 9 ways this can happen and there are 2 possibilities for the remaining 8 holes, each one can either be occupied or not. The number of states is then $9 \cdot 2^8$. Now look at states where the black ball does not occupy a hole. Let k be the number of white balls occupying holes, then k can range from 0 to 8. The number of ways k balls can occupy 9 different holes is $\binom{9}{k}$. The total number of ways the white balls can occupy the holes is then

$$\sum_{k=0}^{8} \binom{9}{k} = 2^9 - 1$$

The total number of states the game can be in is therefore $9 \cdot 2^8 + 2^9 - 1 = 2,815$.

Problem 193. For the previous question, if we now have 2 black balls and 7 white balls, how many states can the game be in?

Answer. There can be 0, 1, or 2 black balls occupying holes, and the number of states for each is:

$$0: \sum_{k=0}^{7} \binom{9}{k} = 2^9 - \binom{9}{8} - \binom{9}{9} = 2^9 - 10$$

$$1: 9 \sum_{k=0}^{7} \binom{8}{k} = 9(2^8 - 1)$$

$$2: \binom{9}{2} \sum_{k=0}^{7} \binom{7}{k} = \frac{9 \cdot 8}{2} 2^7 = 9 \cdot 2^9$$

The total number of states is then
$2^9 - 10 + 9(2^8 - 1) + 9 \cdot 2^9 = 7,405.$

Problem 194. For the previous question, if we now have 1 red ball, 1 green, and 7 white balls, how many states can the game be in?

Answer. With no red or green balls, the number of states is

$$\sum_{k=0}^{7} \binom{9}{k} = 2^9 - \binom{9}{8} - \binom{9}{9} = 2^9 - 10$$

With either a red or green ball, the number of

states is

$$2 \cdot 9 \cdot \sum_{k=0}^{7} \binom{8}{k} = 18(2^8 - 1)$$

With both a red and a green ball, the number of states is

$$2 \cdot \binom{9}{2} \sum_{k=0}^{7} \binom{7}{k} = 9 \cdot 8 \cdot 2^7$$

So the total number of states is $2^9 - 10 + 18(2^8 - 1) + 9 \cdot 8 \cdot 2^7 = 14,308$.

Problem 195. How many ways can you give 27 different books to Alice, Bob, and Cathy so that Alice and Cathy together have twice as many books as Bob?

Answer. Let a, b, and c be the number of books given to Alice, Bob, and Cathy respectively, then $a + b + c = 27$ and $a + c = 2b$. Solving these equations for b and $a + c$ gives, $b = 9$ and $a + c = 18$. 9 books have to be given to Bob, and the remaining 18 are split between Alice and Cathy. Bob can get 9 books in $\binom{27}{9}$ ways, and there are 2 choices for each of the remaining 18 books, so the total number of choices is

$$\binom{27}{9} 2^{18} = 1,228,623,052,800$$

.

Problem 196. For 99 different things, show that the ratio of the number of ways of selecting 70 to the number of ways of selecting 30 is $\frac{3}{7}$.

Answer. The number of ways to select 70 is $\binom{99}{70}$. The number of ways to select 30 is $\binom{99}{30}$. The ratio is

$$\frac{\binom{99}{70}}{\binom{99}{30}} = \frac{99!}{70!29!} \cdot \frac{30!69!}{99!}$$

$$= \frac{30}{70} = \frac{3}{7}$$

.

Problem 197. In how many ways can 8 pizzas be delivered by 4 delivery guys so that they each deliver at least one pizza?

Answer. This question is equivalent to "How many ways can you partition a set of 8 pizzas into 4 distinct nonempty subsets?". In terms of putting balls into bins, this is the number of ways to put 8 distinct balls into 4 distinct bins such that each bin has at least one ball. The number of ways to do this is $4! \cdot S(8,4)$ where $S(8,4)$ is a Stirling number of the second kind defined as

$$S(8,4) = \frac{1}{4!} \sum_{i=0}^{3} (-1)^i \binom{3}{i} (3-i)^8$$

$$= \frac{1}{4!} (4^8 - 4 \cdot 3^8 + 6 \cdot 2^8 - 4)$$

So the number of ways the pizzas can be delivered is

$$4^8 - 4 \cdot 3^8 + 6 \cdot 2^8 - 4 = 40,824$$

Problem 198. How many ways can you select 3 integers from the set $(1, 2, 3, \ldots, 100)$ so that their sum is divisible by 3?

Answer. The sum will be divisible by 3 under 4 conditions:

1. All integers are of the form $x = 3n$, $n = 1, 2, \ldots, 33$

2. All integers are of the form $x = 3n + 1$, $n = 0, 1, 2, \ldots, 33$

3. All integers are of the form $x = 3n + 2$, $n = 0, 1, 2, \ldots, 32$

4. One integer of each of the above forms.

The number of ways to select the integers is then

$$\binom{33}{3} + \binom{34}{3} + \binom{33}{3} + 33 \cdot 34 \cdot 33 = 53,922$$

Problem 199. The ratio of the number of ways to select x out of $2x + 2$ to the number of ways to select x out of $2x - 2$ is $\frac{99}{7}$. What is x?

Answer. The ratio is

$$\frac{\binom{2x+2}{x}}{\binom{2x-2}{x}} = \frac{(2x+2)!}{x!(x+2)!} \cdot \frac{x!(x-2)!}{(2x-2)!}$$

$$= \frac{(2x+2)(2x+1)2x(2x-1)}{(x+2)(x+1)x(x-1)}$$

$$= \frac{4(2x+1)(2x-1)}{(x+2)(x-1)}$$

$$= \frac{99}{7}$$

This simplifies to

$$13x^2 - 99x + 170 = 0$$

Solving for x gives $x = 5$.

Problem 200. If $\binom{n}{3} + \binom{n+2}{3} = \frac{n!}{(n-3)!}$ find n.

Answer. The equation can be written as

$$\frac{n(n-1)(n-2)}{6} + \frac{(n+2)(n+1)n}{6} = n(n-1)(n-2)$$

which simplifies to $2n^2 - 9n + 4 = 0$ then solving for n gives $n = 4$.

Problem 201. If repetitions are allowed, show that the number of ways to select n things out of $m+1$ is the same as the number of ways to select m things out of $n + 1$.

Answer. Another way of saying this is that the number of ways to put n identical balls into $m + 1$ distinct urns is the same as the number of ways to put m identical balls into $n + 1$ distinct urns. In both cases the number is

$$\binom{n + m}{n} = \binom{n + m}{m}$$

Problem 202. If you throw n dice how many different results are possible?

Answer. This is equivalent to asking how many ways n identical balls can be put into 6 distinct urns. The number is $\binom{n+5}{5}$.

Problem 203. A basket contains $2n + r$ apples and $2n - r$ pears. Show that the choice of n apples and n pears is greatest when $r = 0$, with n being constant.

Answer. The apple choices are $\binom{2n+r}{n}$. The pear choices are $\binom{2n-r}{n}$. The choices of n apples and n pears are then

$$\binom{2n+r}{n}\binom{2n-r}{n} = \frac{(2n+r)!(2n-r)!}{(n!)^2(n+r)!(n-r)!}$$

If the choices of n apples and n pears is greatest when $r = 0$ then

$$(\text{choices when } r = x+1) < (\text{choices when } r = x)$$

$$\frac{(2n+x+1)!(2n-x-1)!}{(n!)^2(n+x+1)!(n-x-1)!} < \frac{(2n+x)!(2n-x)!}{(n!)^2(n+x)!(n-x)!}$$

reduces to

$$\frac{2n+x+1}{n+x+1} < \frac{2n-x}{n-x}$$

resulting in $2x+1 > 0$, which is true when $x = 0, 1, 2, \ldots$ So the greatest number of choices occurs when $r = x = 0$.

Another way to solve this problem is that given the expression for the total number of choices, N, solve for r in $dN/dr = 0$, that is, find the maximum of N. This is more easily done by realizing that the maximum of N is the same as the maximum of $\ln N$. This allows us to use Stirling's approximation for the natural logarithm of a factorial, $\ln x! = x \ln x - x$.

If the number of apple choices is N_1 and the number of pear choices is N_2, then the total number of choices is $N = N_1 N_2$. With $\ln N = \ln N_1 + \ln N_2$, we can find $d(\ln N_1)/dr$ and $d(\ln N_2)/dr$ separately, take their sum to get $d(\ln N)/dr$ and solve for r.

$$N_1 = \frac{(2n+r)!}{n!(n+r)!} \quad N_2 = \frac{(2n-r)!}{n!(n-r)!}$$

$$\ln N_1 = \ln(2n+r)! - \ln n! - \ln(n+r)!$$

$$\ln N_2 = \ln(2n-r)! - \ln n! - \ln(n-r)!$$

Using Stirling's approximation then taking d/dr gives

$$\frac{d(\ln N_1)}{dr} \approx \ln(2n + r) - \ln(n + r)$$

$$\frac{d(\ln N_2)}{dr} \approx -\ln(2n - r) + \ln(n - r)$$

Summing the last two equations and setting to zero gives

$$\ln\left(\frac{2n + r}{n + r}\right) = \ln\left(\frac{2n - r}{n - r}\right)$$

A little algebra on the last equation gives $2nr = 0$. Since $n \neq 0$ we must have $r = 0$, so we have shown what we set out to.

Problem 204. How many ways can you select 3 integers out of $3n$ consecutive integers so that their sum is divisible by 3?

Answer. An integer i will have remainder 0, 1, or 2 when divided by 3. Let A, B and C represent sets of integers with remainders 0, 1 and 2 respectively. In a listing of consecutive integers the remainders will be cyclic. For $3n$ consecutive integers this means there will be exactly n integers of each kind so that $|A| = |B| = |C| = n$. To select 3 integers so that their sum is divisible by 3 we can select one each from A, B and C, or we can select all 3 from either A, B, and C. In each case the remainder of the sum will be divisible by 3 so the sum is divisible by 3. The number of ways to select one each from A, B or C is $\binom{n}{3}$ so the total number of ways to select the three integers is

$$n^3 + 3\binom{n}{3} = n^3 + \frac{1}{2}n(n-1)(n-2)$$

Problem 205. There are m parcels of which the first contains n things, the second $2n$ things, the third $3n$ things, and so on. Show that the number of ways of taking n things out of each parcel is $(mn)!/(n!)^m$.

Answer. The parcel number, the number of things in each parcel, and the number of ways to choose n things from each parcel, respectively, are shown in the table below. So by the product rule, the

1	2	3	$\cdots$	m
n	$2n$	$3n$	$\cdots$	mn
1	$\binom{2n}{n}$	$\binom{3n}{n}$	$\cdots$	$\binom{mn}{n}$

number of ways of taking n things out of each parcel is

$$1 \cdot \binom{2n}{n} \cdot \binom{3n}{n} \cdots \binom{mn}{n}$$

$$= 1 \cdot \frac{(2n)!}{n!n!} \cdot \frac{(3n)!}{n!(2n)!} \cdots \frac{(mn)!}{n!((m-1)n)!}$$
$$= \frac{(mn)!}{(n!)^m}$$

Problem 206. A very useful binomial coefficient identity is

$$\binom{n}{k} = \binom{n-1}{k-1} + \binom{n-1}{k}$$

Give a combinatorial proof of this identity.

Answer. A combinatorial proof means that we show that both sides of the equation count the same number of things. The left side of the equation is equal to the number of k element subsets of the set $[n]$. Such a subset will either contain the integer n or not. If it does then we can remove it and what we have left is equivalent to a $k-1$ element subset of the set $[n-1]$ and there are $\binom{n-1}{k-1}$ of those. If it does not then it is equivalent to a k element subset of the set $[n-1]$ and there are $\binom{n-1}{k}$ of those. This proves the identity.

Problem 207. Another useful identity concerning binomial coefficients is Vandermonde's identity

$$\binom{n+m}{k} = \sum_{i=0}^{k} \binom{n}{i}\binom{m}{k-i}$$

Give a combinatorial proof of this identity.

Answer. If we have 2 sets, A and B, of size n and m, respectively, then we can select a total of k elements from these sets by selecting 0 elements from A and k elements from B, or 1 element from A and $k-1$ elements from B, and so on up to k elements from A and 0 elements from B. The righthand side of the identity shows the total number of ways of selecting k elements from the 2 sets. We could also combine the 2 sets into 1 set of size $n+m$. The total number of ways of selecting k elements from the combined set is given by the lefthand side of the identity. This proves the identity.

Problem 208. We have 10 weights that have values $1, 2, ..., 10$ pounds. A number of these weights are selected at random. What is the minimum number that must be selected so that at least one pair of them will sum to exactly 11 pounds?

Answer. This problem can be solved using the pigeonhole principle. The pairs of weights that sum to exactly 11 pounds are $(1, 10)$, $(2, 9)$, $(3, 8)$, $(4, 7)$ and $(5, 6)$. There are 5 pairs and every weight is in one of the pairs. If 6 weights are selected at random then at least two of them must come from the same pair and thus sum to 11.

Problem 209. Given a set of n integer valued weights

show that there is always some subset of them whose weight is divisible by n.

Answer. This problem can be solved using the pigeonhole principle. Let w_1, w_2, w_3,...w_n be the values of the n weights. Form n subsets so that the weight of subset j is $W_j = \Sigma_{i=1}^{j} w_i$. Either one of these W_j is divisible by n or they will all have remainders ranging from 1 to $n - 1$. There are $n - 1$ possible remainders and n W_j's therefore there must be at least two W_j's with the same remainder. Subtracting these two will give us a weight with remainder 0. Subtracting the corresponding subsets will produce the subset whose weight is divisible by n.

Problem 210. Given a dartboard in the shape of an equilateral triangle with sides of length 2. How many darts must be thrown at the board so that at least 2 darts are separated by a distance no greater than 1?

Answer. This problem can be solved using the pigeonhole principle. The equilateral triangle can be partitioned into 4 equilateral triangles with sides of length 1 as shown in figure 5. If 5 darts are thrown at the board then at least 2 of them must land in the same subtriangle so that the distance between them can not be larger than 1.

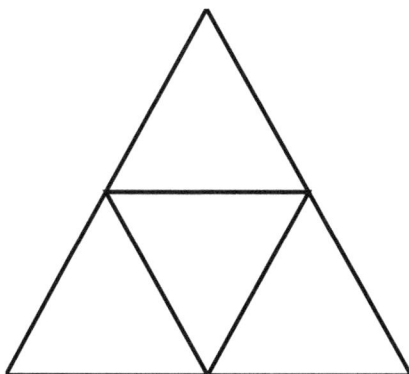

Figure 5: An equilateral triangle of side length 2 partitioned into four equilateral triangles of side length 1.

Problem 211. Sparky is packing for a trip. He has a total of 10 different shirts and 5 different pants. If he can only pack 3 shirts and 2 pants how many different ways can he pack?

Answer.

$$\binom{10}{3}\binom{5}{2} = \frac{10!}{3!(10-3)!} \cdot \frac{5!}{2!(5-2)!} = 1,200 \text{ ways}$$

Problem 212. Alice wants to plant a row of 10 flowers in her front yard. She has 4 kinds of flowers to choose from and she does not want 2 flowers of the same kind next to each other. How many ways can she plant the flowers?

Answer. For the first flower there are 4 choices. For the second flower, all but the type just planted can be chosen, so there are 3 choices. Similarly, for the third flower there are 3 choices, and so on to the tenth. So the total number of ways is

$$4 \cdot 3^9 = 78,732 \text{ ways}$$

Problem 213. Of those $78,732$ choices of the last problem, how many have the first and last flower different?

Answer. As with the last problem, for the first flower there are 4 choices. For the second flower, all but the type just planted can be chosen, so there are 3 choices. Similarly, for the third flower there are 3 choices, and so on to the ninth. But the tenth flower can neither be like its neighbor nor like the first flower, so there are 2 choices for it. The total number of ways is then

$$4 \cdot 3^8 \cdot 2 = 52,488 \text{ ways}$$

Problem 214. Give a combinatorial proof of the identity

$$\frac{1}{2}\binom{2n+2}{n+1} = \binom{2n}{n} + \binom{2n}{n-1}$$

Answer. We will outline the proof. A subset of size $n+1$ can be formed from the set of integers $[2n+2]$ in $\binom{2n+2}{n+1}$ ways. The subset along with the set of integers not chosen is a partition of $[2n+2]$ into

2 sets of size $n + 1$. Since the same set of $n + 1$ integers can be chosen or not, the $\binom{2n+2}{n+1}$ ways of selecting the integers over counts the number of partitions by a factor of 2. The number of unique ways to partition the set $[2n + 2]$ into 2 sets of size $n + 1$ is therefore $\frac{1}{2}\binom{2n+2}{n+1}$. This is the lefthand side of the above identity. In any of the partitions, the integers $2n + 2$ and $2n + 1$ will either appear in different halves of the partition or in the same half. If they are in different halves, delete them, and we are left with a partition of $[2n]$ into 2 sets of size n. If they are in the same half, delete them, and we have a partition of $[2n]$ into 2 sets of size $n - 1$ and n. Doing this for all the partitions of $[2n + 2]$ gives us the identity.

Problem 215. Give a combinatorial proof of the identity

$$\sum_{k=0}^{n} \binom{n}{k} = 2^n$$

Answer. A subset of $[n]$ of size k can be formed in $\binom{n}{k}$ ways. A subset can have $k = 0, 1, \ldots, n$ elements, therefore the lefthand side of the identity gives the total number of subsets of $[n]$. For any subset we can assign the value 1 to those elements of $[n]$ that are in the subset and the value 0 to those elements that are not. Every subset can therefore be represented by an n digit binary number. The total number of subsets must then be equal to the number of n digit binary numbers. Since there are 2 possible values for each digit, the number of n digit binary numbers is 2^n. This proves the identity.

Problem 216. The binomial coefficients, $\binom{n}{k}$, appear in the expansion

$$(x+1)^n = \sum_{k=0}^{n} x^k \binom{n}{k}$$

Show how this is related to the identity in the previous problem.

Answer. Set $x = 1$ and the expansion becomes the identity.

Problem 217. Derive the following identity

$$\sum_{k=1}^{n} k \binom{n}{k} = n2^{n-1}$$

Answer. Take the derivative of the binomial expansion in the previous problem and set $x = 1$. This will produce the above identity.

Problem 218. If $pq + r$ different things are to be divided as equally as possible among p people, in how many ways can it be done with $r < p$?

Answer. The problem amounts to distributing $pq + r$ distinct balls into p distinct bins with as little difference as possible between the numbers in each bin. If $r = 0$ then each of the p bins would get q balls. The number of ways to do this is

$$\frac{(pq)!}{(q!)^p}$$

With $r > 0$ the most uniform way to distribute the balls is to put $q + 1$ into r bins and q balls into $p - r$ bins. The number of ways to do that is

$$\frac{(pq + r)!}{((q + 1)!)^r (q!)^{p-r}} \binom{p}{r}$$

where $\binom{p}{r}$ is the number of ways to select the r bins that will receive $q + 1$ balls.

Problem 219. Prove that the number of ways in which p positive signs and n negative signs may be placed in a row, so that no two negative signs shall be together $(p > n)$, is equal to the number of combinations of $p + 1$ things taken n at a time.

Answer. For p positive signs, we can place a negative sign between any of the positive signs or at either end. So there are $p + 1$ positions to place a negative sign, and we have n positions to fill, making the total number of combinations $\binom{p+1}{n}$.

Problem 220. How many ways can you give 5 bananas to 10 monkeys so that no monkey gets more than 1 banana, assuming the bananas are identical?

Answer. In terms of balls and bins, the monkeys are distinct bins and the bananas are identical balls. The number of ways to place k identical balls into n distinct bins with no more than one ball in each bin is $\binom{n}{k}$. We have $n = 10$ and $k = 5$ so the number of ways is $\binom{10}{5} = 252$.

Problem 221. The number of students in several classes of a school have a constant difference when ordered by increasing size, and a number of prizes equal to this constant difference is to be given to each class, with no student getting more than one prize. Show that if the prizes are all unique then the number of ways of giving them is the same as if all the prizes were given to the largest class.

Answer. If p is the size of the smallest class, d is the constant difference, and k is the number of classes, then the number of ways of giving the prizes is (using the product rule)

$$\prod_{i=0}^{k-1} \frac{(p+id)!}{(p+id-d)!}$$

$$= \frac{p!}{(p-d)!} \cdot \frac{(p+d)!}{p!} \cdot \frac{(p+2d)!}{(p+d)!} \cdots \frac{(p+(k-1)d)!}{(p+(k-2)d)!}$$

$$= \frac{(p + (k-1)d)!}{(p-d)!}$$

This is the number of ways that kd different prizes can be given to a class of $p + (k-1)d$ students. This is what we meant to show.

Problem 222. Show that the number of different throws that can be made with n dice is

$$(1 + n)(1 + \frac{n}{2})(1 + \frac{n}{3})(1 + \frac{n}{4})(1 + \frac{n}{5})$$

Answer. This amounts to asking for how many ways you can place n identical balls into 6 distinct bins with no restrictions. This can be solved using the stars and bars formalism with the formula given by equation 53

$$\binom{n + 6 - 1}{6 - 1} = \binom{n + 5}{5} = \frac{(n+5)!}{5!n!}$$

$$= \frac{(n+5)(n+4)(n+3)(n+2)(n+1)}{5!}$$

Problem 223. In a city of $100,000$ people, what is the minimum number of people that have last names beginning with the same first 3 letters?

Answer. Using the pigeonhole principle, the minimum number of people that have last names beginning with the same first letter are $100,000/26$.

Of those, those that share the first 2 letters are $100,000/26^2$. And of those, those that share the first 3 letters are $100,000/26^3 = 5.69$. Because we want the minimum, we must round this number down to the nearest integer. So the minimum number of people is 5.

Problem 224. A slot machine has 3 spinning reels with the same 20 symbols on each reel, one of the symbols being "7". When the reels stop spinning, a symbol is read off from each of the reels. If at least one of the 3 symbols read off is the number "7", then the player gets a prize. How many ways are there to get a prize?

Answer. The number of possible spins are $20 \cdot 20 \cdot 20 = 8000$. The number of ways to not get a prize are $19 \cdot 19 \cdot 19 = 6859$. So the number of ways to get a prize are the difference $8000 - 6859 = 1141$.

Problem 225. A digital circuit has n digital inputs (value 0 or 1) and produces one digital output. How many such digital circuits are there?

Answer. There are 2^n possible inputs, and for each of those there are 2 possible outputs. So the total number of such digital circuits is 2^{2^n}. A table is shown below for $n = 1$ through 5.

n	possible digital circuits (2^{2^n})
1	4
2	16
3	256
4	65536
5	4.29×10^9

Problem 226. For the last problem, if there were 2 digital outputs, how many such digital circuits would there be?

Answer. There are 2^n possible inputs, and for each of those there are $2^2 = 4$ possible outputs. So the total number of such digital circuits is 4^{2^n}. A table is shown below for $n = 1$ through 5.

n	possible digital circuits (4^{2^n})
1	16
2	256
3	65536
4	4.29×10^9
5	1.84×10^{19}

Problem 227. A website requires users to enter a password. There should be at least one trillion (10^{12}) unique passwords. If only lowercase letters are used, what is the minimum password length required for at least one trillion unique passwords?

Answer. This problem reduces to solving for x in the equation

$$26^x \geq 10^{12}$$

where 26 comes from the number of letters in the English alphabet. Taking the logarithm (any base will do) of both sides, and solving for x we get

$$x \geq \frac{12 \ln 10}{\ln 26} = 8.48$$

Since there are no fractional numbers of letters, we round up, so the minimum password length for at least one trillion unique passwords is 9. Note that the actual number of unique passwords for length 9 is $26^9 = 5.4 \times 10^{12}$, so we've got a comfortable margin of more than a factor of 5.

Problem 228. The number of ways to select 4 things from n different things is 1/6 the number of ways to select 4 things from $2n$ things that have n unique pairs of 2 identical things. Find n.

Answer. If we select 4 things from those $2n$ things, the result must be one of the following cases:

1. All 4 selections are unique.

2. The 4 selections are 2 pair of identical objects.

3. The 4 selections are 1 pair of identical objects, and 2 objects that are different.

For case (1) there are $\binom{n}{4}$ ways to get it. For case (2) there are $\binom{n}{2}$ ways to get it. For case (3) there are $3\binom{n}{3}$ ways to get it because the choice is from 3 unique objects, and one of those must be chosen for the double. So the expression we have to solve for n is

$$\binom{n}{4} = \frac{1}{6}\left[\binom{n}{4} + \binom{n}{2} + 3\binom{n}{3}\right]$$

which reduces to

$$5\binom{n}{4} = \binom{n}{2} + 3\binom{n}{3}$$

$$\frac{5n!}{4!(n-4)!} = \frac{n!}{2(n-2)!} + \frac{3n!}{6(n-3)!}$$

giving the quadratic

$$(5n - 7)(n - 6) = 0$$

Since n must be an integer, our answer is $n = 6$.

Problem 229. In how many ways can a triangle be formed inside a hexagon using 3 of its 6 vertices?

Answer. There are $\binom{6}{3} = 20$ ways to choose 3 of a hexagon's 6 vertices. All the ways are shown in figure 6.

Figure 6: The 20 ways to form a triangle inside a hexagon using 3 of its 6 vertices.

Problem 230. In how many ways can a triangle be
formed inside an octagon using 3 of its 8 vertices?

Answer. There are $\binom{8}{3} = 56$ ways to choose 3 of an
octagon's 8 vertices. All the ways are shown in
figure 7.

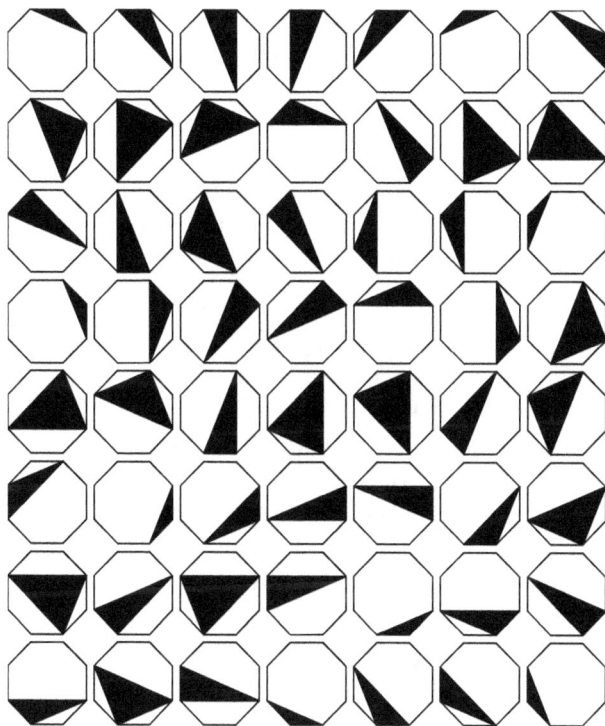

Figure 7: The 56 ways to form a triangle inside an
octagon using 3 of its 8 vertices.

Problem 231. In how many ways can a quadrilateral be formed inside an octagon using 4 of its 8 vertices?

Answer. There are $\binom{8}{4} = 70$ ways to choose 4 of an octagon's 8 vertices. All the ways are shown in figure 8.

Problem 232. How many strings of length n can you construct using k different symbols such that no adjacent symbols in the string are equal?

Answer. There are k choices for the first symbol and $k - 1$ choices for each symbol after that. The number of strings is therefor

$$k(k-1)^{n-1} \tag{71}$$

Problem 233. The strings counted by the formula in the previous problem include strings that do not use all k symbols. Find a formula for the number of strings that use all of the k symbols.

Answer. Let $M(n, k)$ be the number of strings of length n that use exactly k different symbols with no adjacent symbols in the string being equal. We can then write the answer in the previous problem as follows

$$k(k-1)^{n-1} = \sum_{j=2}^{k} \binom{k}{j} M(n, j) \tag{72}$$

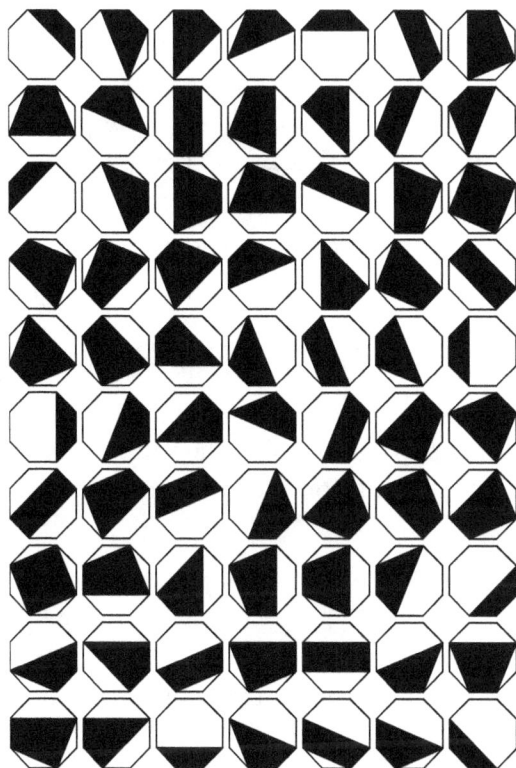

Figure 8: The 70 ways to form a quadrilateral inside an octagon using 4 of its 8 vertices.

Each term in the sum is the number of ways to select j symbols out of k, times the number of strings, $M(n, j)$ that can be constructed with those symbols. From equation 44, which we repeat here

$$k^n = \sum_{j=1}^{k} S(n, j) j! \binom{k}{j} \tag{73}$$

we can write $k(k-1)^{n-1}$ as follows

$$k(k-1)^{n-1} = k \sum_{j=1}^{k-1} S(n-1, j) j! \binom{k-1}{j} \tag{74}$$

Taking the k term inside the summation and changing the summation index, we can write this as

$$k(k-1)^{n-1} = \sum_{j=2}^{k} S(n-1, j-1)(j-1)! \frac{k!}{(j-1)!(k-j)!} \tag{75}$$

Now multiply the summation terms by j/j and we have

$$k(k-1)^{n-1} = \sum_{j=2}^{k} S(n-1, j-1) j! \binom{k}{j} \tag{76}$$

Comparing this with the above equation in terms of $M(n, j)$ we see that $M(n, j) = S(n-1, j-1) j!$. Figure 9 lists some values for $M(n, k)$.

n	2	3	4	5	k 6	7	8	9
2	2							
3	2	6						
4	2	18	24					
5	2	42	144	120				
6	2	90	600	1200	720			
7	2	186	2160	7800	10800	5040		
8	2	378	7224	42000	100800	105840	40320	
9	2	762	23184	204120	756000	1340640	1128960	362880

Figure 9: $M(n, k) = S(n - 1, k - 1)k!$

Problem 234. A farmer wants to plant 6 trees along the edge of his field. He has 2 each of 3 kinds of trees. How many ways can he plant them?

Answer. This amounts to asking for how many ways you can order a multiset of n objects of k different types with a_i objects of type i? The solution is

$$\frac{n!}{a_1! a_2! \cdots a_k!} \qquad (77)$$

In this case $n = 6$, $k = 3$, $a_1 = a_2 = a_3 = 2$, giving

$$\frac{6!}{2! 2! 2!} = 90 \text{ ways.} \qquad (78)$$

All the ways are shown in figure 10.

Problem 235. How many triangles can be formed with every side either 8 or 10 or 12 or 14 millimeters long?

ΦΦΥΥΔΔ ΦΦΥΔΥΔ ΦΦΥΔΔΥ ΦΦΔΥΥΔ ΦΦΔΥΔΥ
ΦΦΔΔΥΥ ΦΥΦΥΔΔ ΦΥΦΔΥΔ ΦΥΦΔΔΥ ΦΥΥΦΔΔ
ΦΥΥΔΦΔ ΦΥΥΔΔΦ ΦΥΔΦΥΔ ΦΥΔΦΔΥ ΦΥΔΥΦΔ
ΦΥΔΥΔΦ ΦΥΔΔΦΥ ΦΥΔΔΥΦ ΦΔΦΥΥΔ ΦΔΦΥΔΥ
ΦΔΦΔΥΥ ΦΔΥΦΥΔ ΦΔΥΦΔΥ ΦΔΥΥΦΔ ΦΔΥΥΔΦ
ΦΔΥΔΦΥ ΦΔΥΔΥΦ ΦΔΔΦΥΥ ΦΔΔΥΦΥ ΦΔΔΥΥΦ
ΥΦΦΥΔΔ ΥΦΦΔΥΔ ΥΦΦΔΔΥ ΥΦΥΦΔΔ ΥΦΥΔΦΔ
ΥΦΥΔΔΦ ΥΦΔΦΥΔ ΥΦΔΦΔΥ ΥΦΔΥΦΔ ΥΦΔΥΔΦ
ΥΦΔΔΦΥ ΥΦΔΔΥΦ ΥΥΦΦΔΔ ΥΥΦΔΦΔ ΥΥΦΔΔΦ
ΥΥΔΦΦΔ ΥΥΔΦΔΦ ΥΥΔΔΦΦ ΥΔΦΦΥΔ ΥΔΦΦΔΥ
ΥΔΦΥΦΔ ΥΔΦΥΔΦ ΥΔΦΔΦΥ ΥΔΦΔΥΦ ΥΔΥΦΦΔ
ΥΔΥΦΔΦ ΥΔΥΔΦΦ ΥΔΔΦΦΥ ΥΔΔΦΥΦ ΥΔΔΥΦΦ
ΔΦΦΥΥΔ ΔΦΦΥΔΥ ΔΦΦΔΥΥ ΔΦΥΦΥΔ ΔΦΥΦΔΥ
ΔΦΥΥΦΔ ΔΦΥΥΔΦ ΔΦΥΔΦΥ ΔΦΥΔΥΦ ΔΦΔΦΥΥ
ΔΦΔΥΦΥ ΔΦΔΥΥΦ ΔΥΦΦΥΔ ΔΥΦΦΔΥ ΔΥΦΥΦΔ
ΔΥΦΥΔΦ ΔΥΦΔΦΥ ΔΥΦΔΥΦ ΔΥΥΦΦΔ ΔΥΥΦΔΦ
ΔΥΥΔΦΦ ΔΥΔΦΦΥ ΔΥΔΦΥΦ ΔΥΔΥΦΦ ΔΔΦΦΥΥ
ΔΔΦΥΦΥ ΔΔΦΥΥΦ ΔΔΥΦΦΥ ΔΔΥΦΥΦ ΔΔΥΥΦΦ

Figure 10: The 90 ways to plant 6 trees with 2 each of 3 kinds of trees.

Answer. This amounts to asking for how many multisets of size n you can create by sampling with replacement from a set of size k? The solution is

$$\binom{n + k - 1}{k - 1} \tag{79}$$

Here the multisets are triangle side lengths so $n = 3$. The set we are sampling from is the 4 side lengths so $k = 4$. The number of triangles is then:

$$\binom{3 + 4 - 1}{4 - 1} = \binom{6}{3} = 20 \text{ ways.} \tag{80}$$

All the ways are shown in figure 11.

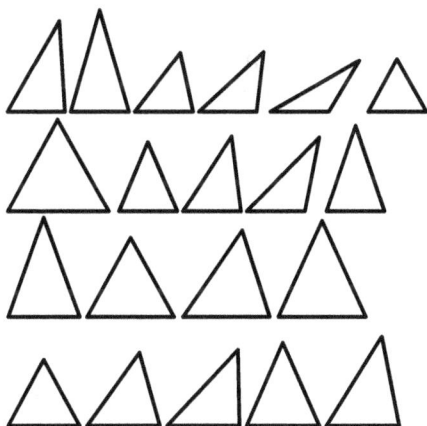

Figure 11: The 20 triangles that can be formed with every side either 8 or 10 or 12 or 14 millimeters long.

Problem 236. How many 3 and 4 digit even numbers are there where all the digits are different? The numbers may not begin with 0.

Answer. For a number to be even it must end with $0, 2, 4, 6, 8$. To get the number of two digit even numbers there are 9 digits that can be added onto each of the 5 one digit even numbers, but 4 of these will begin with a 0. If we want only the numbers that don't begin with a 0 the number of even two digit numbers is $9 \cdot 5 - 4 = 41$. To get a three digit number you can add eight digits to each of the two digit numbers. This will result in some numbers that begin with a 0 but the first two digits of those numbers can be interchanged and the result will still be a number with the required properties. Therefore the number of three digit numbers is $8 \cdot 41 = 328$. Likewise to get four digit numbers we can add seven digits to each of the three digit numbers for a total of $7 \cdot 8 \cdot 41 = 2296$ four digit numbers. Below are all 328 of the three digit numbers.

```
102 104 106 108 120 124 126 128 130 132 134 136 138 140 142 146 148 150 152
154 156 158 160 162 164 168 170 172 174 176 178 180 182 184 186 190 192 194
196 198 204 206 208 210 214 216 218 230 234 236 238 240 246 248 250 254 256
258 260 264 268 270 274 276 278 280 284 286 290 294 296 298 302 304 306 308
310 312 314 316 318 320 324 326 328 340 342 346 348 350 352 354 356 358 360
362 364 368 370 372 374 376 378 380 382 384 386 390 392 394 396 398 402 406
408 410 412 416 418 420 426 428 430 432 436 438 450 452 456 458 460 462 468
470 472 476 478 480 482 486 490 492 496 498 502 504 506 508 510 512 514 516
518 520 524 526 528 530 532 534 536 538 540 542 546 548 560 562 564 568 570
572 574 576 578 580 582 584 586 590 592 594 596 598 602 604 608 610 612 614
618 620 624 628 630 632 634 638 640 642 648 650 652 654 658 670 672 674 678
680 682 684 690 692 694 698 702 704 706 708 710 712 714 716 718 720 724 726
728 730 732 734 736 738 740 742 746 748 750 752 754 756 758 760 762 764 768
780 782 784 786 790 792 794 796 798 802 804 806 810 812 814 816 820 824 826
830 832 834 836 840 842 846 850 852 854 856 860 862 864 870 872 874 876 890
892 894 896 902 904 906 908 910 912 914 916 918 920 924 926 928 930 932 934
936 938 940 942 946 948 950 952 954 956 958 960 962 964 968 970 972 974 976
978 980 982 984 986
```

Problem 237. A town has an average of 10 automobile accidents in a week. In how many ways can those accidents be distributed over the days of the week? Look at the case where we distinguish between different days and where we don't.

Answer. This is a balls into boxes distribution problem where the accidents are the balls and the days are the boxes. Since we are talking about averages, we can assume the accidents are indistinguishable. If we want to distinguish between days of the week then the number of ways is $\binom{10+7-1}{7-1} = \binom{16}{6} = 8008$. If we make no distinction between days of the week, then the number of ways is equal to the number of partitions of 10 into 7 or fewer parts.

$$\sum_{k=1}^{7} p(10, k) = 1 + 5 + 8 + 9 + 7 + 5 + 3 = 38$$

Problem 238. Rework the previous problem where every day has at least one accident.

Answer. If we distinguish between days then we assign one accident to each day and are left with 3 to distribute over the 7 days in $\binom{3+7-1}{7-1} = \binom{9}{6} = 84$. If we don't distinguish between the days then the number of ways is the number of partitions of 10 into exactly 7 parts, $p(10, 7) = 3$. The partitions are $\{1, 1, 1, 1, 1, 1, 4\}$, $\{1, 1, 1, 1, 1, 2, 3\}$, $\{1, 1, 1, 1, 2, 2, 2\}$.

Problem 239. In how many ways can you make change for a dollar using nickels, dimes and quarters? For international readers, a dollar is 100 cents, a quarter is 25 cents, and dimes and nickels are 10 and 5 cents respectively.

Answer. We can divide the ways according to how many quarters are used. Let n_i be the number of ways to make change using i quarters. With no quarters, we are using only dimes and nickels, and both of them divide evenly into a dollar. We can therefor use $0, 1, \ldots 10$ dimes, with the nickels making up the rest of the amount. So we have $n_0 = 11$ ways to use no quarters. With one quarter, an odd number of nickels will have to be used. The number of nickels can be $1, 3, 5, \ldots 15$ there are 8 possible values for the number of nickels therefor $n_1 = 8$. For two quarters there can be between 0 and 5 dimes, therefor $n_2 = 6$. For three quarters there can be 1, 3 or 5 nickels, therefor $n_3 = 3$. For four quarters there are no nickels or dimes, therefor $n_4 = 1$. So the number of ways is $n_0 + n_1 + n_2 + n_3 + n_4 = 11 + 8 + 6 + 3 + 1 = 29$.

There is a simpler and more general way to solve this problem using generating functions. For this problem we construct the function

$$g(x) = \frac{1}{(1 - x^5)(1 - x^{10})(1 - x^{25})}$$

Using any computer algebra system we can find

the power series expansion of this function. We get

$$g(x) = 1+x^5+2\,x^{10}+2\,x^{15}+3\,x^{20}+4\,x^{25}+5\,x^{30}+$$
$$6\,x^{35}+7\,x^{40}+8\,x^{45}+10\,x^{50}+11\,x^{55}+13\,x^{60}+14\,x^{65}+$$
$$16\,x^{70}+18\,x^{75}+20\,x^{80}+22\,x^{85}+24\,x^{90}+26\,x^{95}+$$
$$29\,x^{100}+\cdots$$

So the coefficient of x^{100} is 29 which is the answer we found above. The coefficient of x^n in the expansion is the number of ways to create n cents using nickels, dimes and quarters.

This is a very simple example of using generating functions to solve a problem. We can generalize this example by asking for the number of ways to partition the number n using the k parts $p_1, p_2, \ldots, p_k$. The generating function in this case is

$$g(x) = \frac{1}{\prod_{i=1}^{k}(1 - x^{p_i})}$$

The coefficient of x^n in the expansion of $g(x)$ will give us the answer.

Problem 240. Solve the previous problem using pennies (1 cent) in addition to the nickels, dimes and quarters.

Answer. As in the previous problem we could look at the various ways the coins can be combined to make up one dollar but this becomes exceedingly cumbersome. Instead we'll solve it using generating functions. In this case the generating function is

$$g(x) = \frac{1}{(1-x)(1-x^5)(1-x^{10})(1-x^{25})}$$

If you do a power series expansion of $g(x)$ the coefficient of x^{100} is 242. This is the number of ways to make change for a dollar using pennies, nickels, dimes and quarters.

Problem 241. In American football, a team can score in the following ways:

- 6 points is a touchdown
- 7 points is a touchdown with 1 point conversion
- 8 points is a touchdown with 2 point conversion
- 3 points is a field goal
- 2 points is a safety

The final score in Superbowl 2024 was Kansas City 25 to San Francisco 22. If you did not watch the game, in how many ways could such a score have occurred, not taking into account the order in which the points were scored?

Answer. We will use the method of generating functions we used in the previous two problems. The generating function is

$$g(x) = \frac{1}{(1 - x^2)(1 - x^3)(1 - x^8)(1 - x^7)(1 - x^6)}$$

Notice that there is a $1 - x^p$ term for all the ways points can be scored, $p = 2, 3, 8, 7, 6$. To find the number of ways to score 25 and 22 we look at the coefficients of x^{25} and x^{22} in the power series expansion of $g(x)$. The coefficient of x^{25} is 38 and the coefficient of x^{22} is 30, so the total number of ways the game could have been played is $38 \cdot 30 = 1140$.

Problem 242. In the previous problem, if we exclude safeties and touchdowns with 2 point conversions, which are very rare, how many ways are there to get a final score of 25 to 22, not taking into account the order in which the points were scored?

Answer. We delete the $1 - x^2$ and $1 - x^8$ terms in the generating function of the previous problem so we have

$$g(x) = \frac{1}{(1 - x^3)(1 - x^7)(1 - x^6)}$$

Doing a power series expansion, the coefficient of x^{25} is 4 and the coefficient of x^{22} is 3, so the total number of ways the score in the game could have occurred is $4 \cdot 3 = 12$. The four ways to get 25 points is $(3, 3, 3, 3, 3, 3, 7)$, $(3, 3, 3, 3, 6, 7)$, $(3, 3, 6, 6, 7)$, $(6, 6, 6, 7)$. The three ways to get 22 points is $(3, 3, 3, 3, 3, 7)$, $(3, 3, 3, 6, 7)$, $(3, 6, 6, 7)$.

Problem 243. The actual way that Kansas City scored 25 points and San Francisco scored 22 points is $(3, 3, 3, 3, 6, 7)$ and $(3, 3, 3, 6, 7)$ respectively. This however does not take into account the order in which the points were scored. Considering all possible orders, in how many ways could these points have occurred?

Answer. The number of ways the points $(3, 3, 3, 3, 6, 7)$ could have occurred is $6!/4! = 30$. The number of ways the points $(3, 3, 3, 6, 7)$ could have occurred is $5!/3! = 20$. So the total number of ways the points could have occurred in the game is $30 \cdot 20 = 600$.

Problem 244. How many different rectangular parallelepipeds can be made where the length of each edge is a whole number of centimeters from 1 to 10? The lengths need not all be different.

Answer. This amounts to asking for how many multisets of size n you can create by sampling with replacement from a set of size k? In this case

$n = 3$, $k = 10$, so the number of different rectangular parallelepipeds is

$$\binom{n + k - 1}{k - 1} = \binom{3 + 10 - 1}{10 - 1} = \binom{12}{9} = 220$$

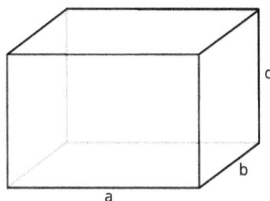

Figure 12: A rectangular parallelepiped of sides (a,b,c).

Problem 245. DNA is a molecule composed of a linear chain of smaller molecules called nucleotides. There are four kinds of nucleotides represented by the letters ACTG. A DNA chain can be represented as a string of these four letters. The sequence of letters in a DNA chain encode for a sequence of amino acids. In the simplest encoding scheme a fixed number of contiguous DNA letters, called a codon, encodes an amino acid. Given that there are 20 different amino acids, what is the minimum number of DNA letters required to encode an amino acid?

Answer. With two of four different kinds of letters only $4^2 = 16$ different things can be encoded.

With three letters we can encode $4^3 = 64$ different things. Therefor we need at least 3 DNA letters to encode the amino acids.

Problem 246. In the previous problem we discovered that a DNA codon needs at least three letters to encode the 20 amino acids. The encoding is however very redundant since three letters can encode for 64 different things. In 1954, just two years after the discovery of the structure of DNA, the physicist George Gamow suggested that the order of the letters in a codon does not matter. All permutations of the letters encode for the same amino acid. If this is true how many amino acids can be encoded?

Answer. In this case the positions of letters in the codon become indistinguishable so we are only interested in how many of each type of letter are in the codon. The problem is equivalent to counting how many ways 3 indistinguishable balls (letters of the codon) can be distributed into 4 bins (DNA letters ACTG) The number of ways is $\binom{3+4-1}{4-1} = \binom{6}{3} = 20$. This exactly matches the number of amino acids that need to be encoded so it looks like Gamow may be right. Later research however showed that this is not correct. More than one codon does encode the same amino acid and there are codons that mark the beginning and end of an amino acid chain.

Problem 247. If there are n points in a plane with $p > 2$ of them in a straight line, how many triangles can be formed using any of the n points for vertices?

Answer. If there are no more than 2 points in a straight line, then the number of triangles that can be formed from those points is $\binom{n}{3}$. But if $p \geq 3$, then that number is reduced by $\binom{p}{3}$, leaving the number of triangles formed to be $\binom{n}{3} - \binom{p}{3}$. An example with $n = 5$, $p = 3$ is shown in figure 13, where the total number of triangles is $\binom{5}{3} - \binom{3}{3} = 9$, whose vertices are (1,2,5), (1,3,5), (1,2,3), (1,2,4), (1,3,4), (1,4,5), (2,3,5), (2,4,5), (3,4,5).

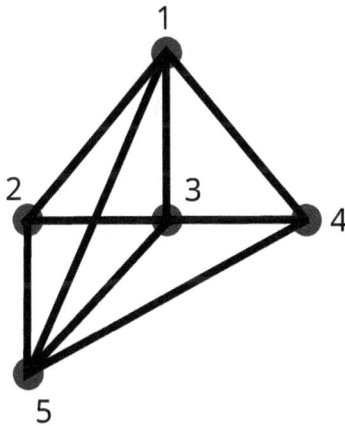

Figure 13: 9 triangles formed by 5 points, with 3 in a straight line.

Problem 248. Quantum mechanical systems have discrete energy levels and each level has a number of states that a particle with that energy can be in. There are two fundamental kinds of particles called fermions and bosons. The electrons, protons, and neutrons that make up atomic matter are all examples of fermions. Examples of bosons are photons and atomic nuclei where the total number of protons and neutrons is even. Fermions obey the Pauli exclusion principle which simply means that there can be no more than one fermion in a state. Bosons on the other hand are more gregarious and have no limits on how many can be in a state. If there are m states and n indistinguishable particles how many ways can the particles occupy the states in the two cases where they are all fermions or all bosons.

Answer. If the particles are all fermions then, since there can be no more than one per state, we must have $m \geq n$ and the number of ways to distribute the particles is $\binom{m}{n}$. If the particles are all bosons then any number can go into a state and there can be more particles than states so the number of ways to distribute them is $\binom{n+m-1}{m-1}$.

Problem 249. An epidemiologist notices that on a street with 12 houses the people in 3 contiguous houses have all contracted a virus. In how many ways can the people in 3 out of 12 houses have a virus? In how many ways are the 3 houses contiguous? Do these numbers indicate that the virus may be spreading between houses?

Answer. The number of ways people in 3 of 12 houses can have a virus is $\binom{12}{3} = 220$. The number of ways in which the houses are contiguous is 10 which is a small fraction of 220. It looks like the virus may be spreading between houses.

Problem 250. In how many ways can 4 black balls, 4 white balls, and 4 red balls be put into 6 pockets where one or more can remain empty?

Answer. This problem can be broken down into three identical smaller problems: how many ways can I put 4 black balls into 6 pockets where some pockets can remain empty? And similarly for the white and red balls. Viewing this as a stars and bars problem (see the Balls in Boxes section), the pockets can be represented by bars, of which we need $6 - 1 = 5$, and the 4 balls of the same color are represented by 4 stars. The number of ways for the 4 balls to be put into the 6 pockets is then $\binom{5+4}{5} = \binom{9}{5} = 126$. Since we have 3 colors of 4 balls each, the total number of ways is then $126^3 = 2,000,376$.

Problem 251. In how many ways can you arrange 20

books on five shelves. Assume each shelf is long enough to hold 20 books?

Answer. Line up the 20 books along with 4 dividers to separate the books into 5 groups corresponding to the 5 shelves. There are 24! ways the books and dividers can be arranged but the dividers are indistinguishable so the number of unique arrangements is $24!/4! = 23! =$
$25, 852, 016, 738, 884, 976, 640, 000$.

Problem 252. Solve the previous problem with the added condition that each shelf must hold at least one book.

Answer. Begin by selecting one book for each shelf. There are 20 ways to select a book for the first shelf, 19 ways for the second shelf and so on. The number of ways to select the first book for each shelf is therefore $20!/15!$. The remaining 15 books can be arranged, as in the previous problem, in $19!/4!$ ways. The total number of ways to arrange the books is then $(20!/15!)(19!/4!) =$
$9, 429, 928, 183, 692, 656, 640, 000$.

Problem 253. In how many ways can you arrange 20 books on five shelves with exactly 4 books on each shelf.

Answer. For every arrangement of the 20 books we put the first four on the first shelf, the second four on the second shelf and so on until all five shelves have four books. There are no books left over so the total number of arrangements is the number of ways we can arrange 20 books or $20! = 2,432,902,008,176,640,000$.

Problem 254. In how many ways can a person wear five rings on two hands? The rings can not be worn on the thumbs and no more than one ring is allowed on a finger.

Answer. Excluding the thumbs, there are 8 fingers that can have rings. We can select the 5 fingers that will receive a ring in $\binom{8}{5} = 56$ ways. Assuming the rings are distinct, they can be placed on the 5 fingers in $5! = 120$ ways. The total number of ways to wear the rings is then $56 \cdot 120 = 6,720$.

Problem 255. Solve the previous problem with the condition that the five rings consist of 3 identical rings of one kind and 2 identical rings of another kind.

Answer. We still have $\binom{8}{5} = 56$ ways to select the fingers but the number of ways to place the rings on the fingers is now $\frac{5!}{2!3!} = 10$. The total number of ways to wear the rings is then $56 \cdot 10 = 560$.

Problem 256. A society of fabulists with 25 members is having an election to select a president. There are 3 candidates. Assuming the votes are anonymous and each of the candidates votes for themselves, how many possible election outcomes are there?

Answer. We want to find the number of ways to put 25 anonymous votes into 3 distinct bins with at least one vote in each bin. With one vote in each bin there are 22 remaining votes that can be distributed among the 3 bins in $\binom{24}{2} = 276$ ways.

Problem 257. A bookbinder has 12 different books to bind in red, green or black cloth. In how many different ways can she bind them, binding at least one in each color?

Answer. Begin by counting how many ways the set of 12 books can be divided into 3 subsets. The number of ways to do this is given by the Stirling number of the second kind $S(12, 3)$. From equation 41 we have $S(n, 3) = \frac{1}{3!}(3^n - 3 \cdot 2^n + 3)$ giving $S(12, 3) = 86,526$. For each of these divisions into 3 subsets we can assign 3 colors to the subsets in $3! = 6$ ways. So the total number of ways to bind the books is $3!S(12, 3) = 519,156$.

Problem 258. In how many ways can 26 different letters be made into six words, each letter being used once and only once?

Answer. Each word must have at least one letter. We can select these letters in 26!/20! ways. For the remaining 20 letters we add 5 partitions to delineate the words. The number of unique permutations of 20 letters and 5 partitions is equal to 25!/5!. The number of ways to construct six words is then equal to

$$\frac{26!}{20!}\frac{25!}{5!} = 2.14268753296565574185779 \cdot 10^{31}$$

If the order of the words is not important then we divide this by 6! to get
$2.975954906896744085913 \cdot 10^{28}$ ways.

Problem 259. Show that the number of ways to partition the integer $2n$ into n parts is equal to the total number of ways to partition the integer n.

Answer. Begin by setting each of the n parts of $2n$ equal to 1. To complete the partition a total of n must be added to these parts. The way to do this is equal to the total number of ways to partition n.

Problem 260. In how many ways can you give 30 apples to 8 horses without giving less than 2 apples to any one horse?

Answer. First let's give the minimum requirement of 2 apples to each horse. This consumes 16 apples of 30, leaving 14 apples. The problem remaining then, in terms of balls and boxes, is how many ways can I distribute 14 identical balls among 8 distinct boxes with no restrictions. The solution with $n = 14$, $k = 8$ is

$$\binom{n+k-1}{k-1} = \binom{14+8-1}{8-1} = \binom{21}{7}$$
$$= 116,280 \text{ ways.}$$

Problem 261. A group of n people wants to select one of their members to represent them. If there's a roll call and every member votes, how many outcomes are there? If the votes are anonymous, how many outcomes are there?

Answer. In the first case we have n distinguishable voters and each has n distinguishable choices, so the number of outcomes is n^n. In the second case we have n indistinguishable voters each with n distinguishable choices. The number of ways to distribute n indistinguishable voters into n distinguishable choices is $\binom{2n-1}{n-1}$.

Problem 262. Given m black balls and n white balls, arrange them in a line so that there are $2r - 1$ places where a group of one or more black balls

contacts a group of one or more white balls. How
many ways can this be done?

Answer. The $2r - 1$ contact points means there must
be $2r$ groups of balls. Label the groups from 1 to
$2r$. Put one color in the odd numbered groups,
and the other color in the even numbered groups.
There are 2 ways this can be done. The m black
balls can be distributed into r groups so that
there is at least one ball in each group in $\binom{m-1}{r-1}$
ways. Likewise the number of ways to distribute
the n white balls into r groups, with at least one
ball in each group, is $\binom{n-1}{r-1}$. The total number
of ways to distribute the white and black balls is
$2\binom{m-1}{r-1}\binom{n-1}{r-1}$.

Problem 263. For the previous problem, what are
the number of ways if there are $2r$ contact points?

Answer. The $2r$ contact points means there must be
$2r+1$ groups of balls. Label the groups from 1 to
$2r+1$. Again, there are 2 ways to put one color in
the odd numbered groups, and the other color in
the even numbered groups. But this time there
are $r+1$ odd numbered groups, and r even numbered groups. Using the sum rule to account for
the different odd and even numbers of groups, the
total number of ways is $\binom{m-1}{r}\binom{n-1}{r-1} + \binom{m-1}{r-1}\binom{n-1}{r}$.
This reduces to $\binom{m-1}{r-1}\binom{n-1}{r-1}\frac{m+n-2r}{r}$. Note that
the factor $\frac{m+n-2r}{r}$ must be positive, which means

$m + n > 2r$. So if $r = 1$ then the sum $m + n$ must be at least 3 which is indeed the minimum number of balls for 2 contact points.

Further Reading

1. *A Course in Enumeration*, Martin Aigner

2. *102 Combinatorial Problems*, Titu Andreescu

3. *Integer Partitions*, George E. Andrews

4. *Schaum's Outline of Theory and Problems of Combinatorics*, V.K. Balakrishnan

5. *Proofs that Really Count: The Art of Combinatorial Proof*, Benjamin and Quinn

6. *The roots of combinatorics*, N. L. Biggs, Historia Mathematica, Vol 6, 1979, p109-136

7. *A Combinatorial Miscellany*, Bjorner and Stanley

8. *Selected combinatorial problems of computational biology*, Jacek Blazewicz, Piotr Formanowicz, Marta Kasprzak, European Journal of Operational Research, Vol 161, No 3, 2005, p585-597

9. *Combinatorics Through Guided Discovery*, Kenneth P. Bogart, 2004

10. *Combinatorics of Permutations*, Miklos Bona

11. *A Walk Through Combinatorics: An Introduction to Enumeration and Graph Theory*, Miklos Bona

12. *The stable marriage problem and sudoku*, Borodin, et al., The College Mathematics Journal, Oct 19, 2023

13. *Close Encounters with the Stirling Numbers of the Second Kind*, Khristo N. Boyadzhiev, Mathematics Magazine, Vol 85, No 4, p252-266, 2012

14. *Introductory Combinatorics*, Richard A. Brualdi

15. *Advanced combinatorics; the art of finite and infinite expansions*, Louis Comtet

16. *A combinatorial problem*, N. G. de Bruijn, Proceedings of the Section of Sciences of the Koninklijke Nederlandse Akademie van Wetenschappen te Amsterdam, Vol 49, No 7, p758-764, 1946

17. *Analytic Combinatorics*, Flajolet and Sedgewick

18. *Polyominoes : puzzles, patterns, problems, and packings*, Solomon W. Golomb, 1996

19. *Concrete mathematics : a foundation for computer science*, Graham, Knuth, Patashnik, 2nd ed, 1998

20. *Combinatorics and Graph Theory*, Harris, Hirst, Mossinghoff, 2nd ed, 2008

21. *The pigeonhole principle, two centuries before Dirichlet*, Albrecht Heeffer and Benoit Rittaud, The

Mathematical Intelligencer, Vol 36, No 2, p27-29, 2014

22. *Combinatorics of compositions and words*, Silvia Heubach and Toufik Mansour, 2010

23. *Adventures in group theory : Rubik's cube, Merlin's machine, and other mathematical toys*, David Joyner

24. *Applied Combinatorics*, Mitchel T. Keller and William T. Trotter, 2017

25. *Combinatorial algorithms : generation, enumeration, and search*, Donald L. Kreher and Douglas R. Stinson, 1999

26. *Combinatorial Problems and Exercises*, Laszlo Lovasz

27. *Discrete mathematics : elementary and beyond*, Laszlo Lovasz

28. *Combinatorics : a guided tour*, David R. Mazur

29. *Combinatorics*, Russell Merris

30. *Combinatorial algorithms for computers and calculators*, Albert Nijenhuis and Herbert S. Wilf, 1978

31. *Let's expand Rota's twelvefold way for counting partitions!*, Robert A. Proctor, Apr 2007

Acknowledgments

In ordinary life we hardly realize that we receive a great deal more than we give, and that it is only with gratitude that life becomes rich. It is very easy to overestimate the importance of our own achievements in comparison with what we owe to others.

Dietrich Bonhoeffer, letter to parents from prison, Sept. 13, 1943

We'd like to thank our parents, Istvan and Anna Hollos, for helping us in many ways.

We thank the makers and maintainers of all the software we've used in the production of this book, including: the Emacs text editor, the LaTex typesetting system, Inkscape, Evince document viewer, Maxima computer algebra system, gcc, bash shell, and the Linux operating system.

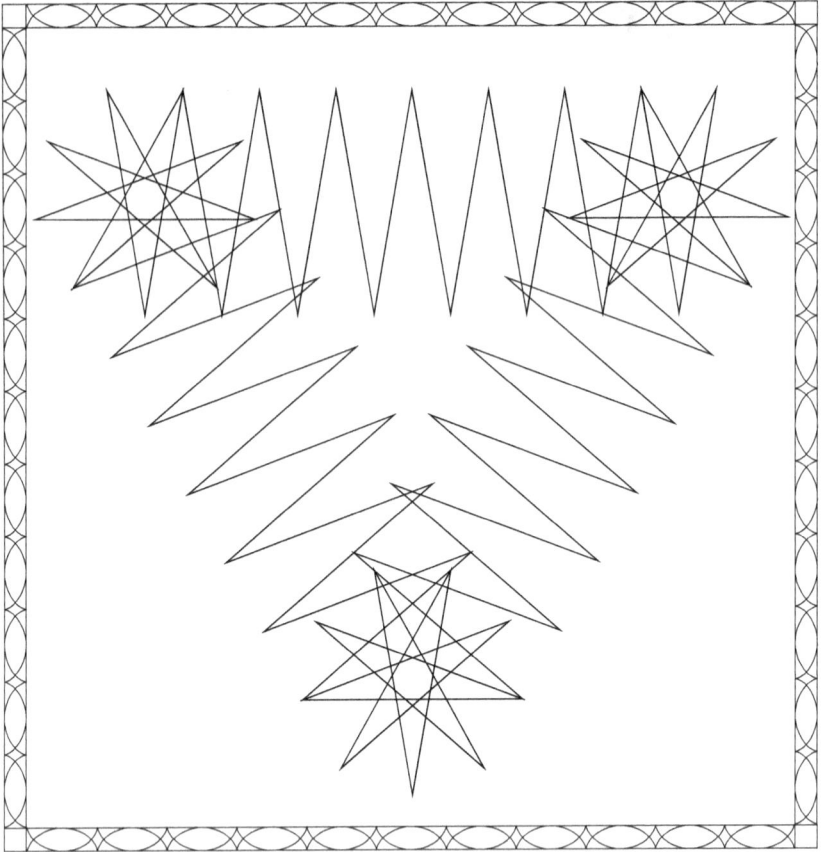

That side of our existence whose direction is towards
the infinite seeks not wealth, but freedom and joy.
Rabindranath Tagore

Stefan Hollos and J. Richard Hollos are physicists by training, and enjoy anything related to physics, engineering and math. They are the authors of

- Random Walks in Electrical Networks

- Creating Noise, second edition

- Engineer's Notebook on Inductor and Transformer Circuits: Problems, Solutions and Simulations

- The Enigma of the Crookes Radiometer

- Passive Butterworth Filter Cookbook

- Nell: An SVG Drawing Language

- Coin Tossing: The Hydrogen Atom of Probability

- Creating Melodies

- Hexagonal Tilings and Patterns

- Combinatorics II Problems and Solutions: Counting Patterns

- Information Theory: A Concise Introduction

- Recursive Digital Filters: A Concise Guide

- Art of Pi

- Creating Noise

- Art of the Golden Ratio

- Creating Rhythms

- Pattern Generation for Computational Art

- Finite Automata and Regular Expressions: Problems and Solutions

- Probability Problems and Solutions

- Combinatorics Problems and Solutions

- The Coin Toss: Probabilities and Patterns

- Pairs Trading: A Bayesian Example

- Simple Trading Strategies That Work

- Bet Smart: The Kelly System for Gambling and Investing

- Signals from the Subatomic World: How to Build a Proton Precession Magnetometer

They are brothers and business partners at Exstrom Laboratories LLC in Longmont, Colorado. The websites for their work is exstrom.com and abrazol.com where you can sign up for their newsletter.

Thank You

Thank you for buying this book.

Sign up for the Abrazol Publishing Newsletter and receive news on updates, new books, and special offers. Just go to

https://www.abrazol.com/

and enter your email address.

The web page for this book is also located at abrazol.com.

www.ingramcontent.com/pod-product-compliance
Lightning Source LLC
Chambersburg PA
CBHW070524200326
41519CB00013B/2925